An Introduction to Electronic Warfare; from the First Jamming to Machine Learning Techniques

RIVER PUBLISHERS SERIES IN SIGNAL, IMAGE AND SPEECH PROCESSING

Series Editors:

MARCELO SAMPAIO DE ALENCAR
Universidade Federal da Bahia UFBA, Brasil

MONCEF GABBOUJ
Tampere University of Technology, Finland

THANOS STOURAITIS
University of Patras, Greece
and
Khalifa University, UAE

The "River Publishers Series in Signal, Image and Speech Processing" is a series of comprehensive academic and professional books which focus on all aspects of the theory and practice of signal processing. Books published in the series include research monographs, edited volumes, handbooks and textbooks. The books provide professionals, researchers, educators, and advanced students in the field with an invaluable insight into the latest research and developments.

Topics covered in the series include, but are by no means restricted to the following:

- Signal Processing Systems
- Digital Signal Processing
- Image Processing
- Signal Theory
- Stochastic Processes
- Detection and Estimation
- Pattern Recognition
- Optical Signal Processing
- Multi-dimensional Signal Processing
- Communication Signal Processing
- Biomedical Signal Processing
- Acoustic and Vibration Signal Processing
- Data Processing
- Remote Sensing
- Signal Processing Technology
- Speech Processing
- Radar Signal Processing

For a list of other books in this series, visit www.riverpublishers.com

An Introduction to Electronic Warfare; from the First Jamming to Machine Learning Techniques

Chi-Hao Cheng

Miami University, USA

James Tsui

US Air Force Research Laboratory (retired), USA

Routledge
Taylor & Francis Group

LONDON AND NEW YORK

Published 2021 by River Publishers
River Publishers
Alsbjergvej 10, 9260 Gistrup, Denmark
www.riverpublishers.com

Distributed exclusively by Routledge
4 Park Square, Milton Park, Abingdon, Oxon OX14 4RN
605 Third Avenue, New York, NY 10017, USA

An Introduction to Electronic Warfare; from the First Jamming to Machine Learning Techniques / by Chi-Hao Cheng, James Tsui.

Routledge is an imprint of the Taylor & Francis Group, an informa business

ISBN 978-87-7022-435-2 (print)

While every effort is made to provide dependable information, the publisher, authors, and editors cannot be held responsible for any errors or omissions.

In memory of Dr. James B. Y. Tsui (1935–2019)

Contents

Preface

The long-lasting duel between radar and electronic warfare (EW) systems is a thought-provoking subject, yet known mostly only to professionals specializing in this field. Nevertheless, technologies developed for radar and EW systems have been applied in numerous areas such as communications, traffic control, cookery (that is right, the microwave oven was accidently invented by a radar engineer), etc., thus shaping our everyday lives. Moreover, the history about how radar and EW engineers have tried to outsmart each other can be an exciting read. The main purpose of this book is to present fundamental concepts of radar and EW systems in a way that is accessible to both technical and non-technical readers. Throughout the book, historical events are used to illustrate how radar/EW technologies were developed and evolved. The authors have intentionally limited the use of equations; however, materials presented in this book are intended to be self-contained and the authors' wish is that, regardless of their educational background, readers can obtain interesting information otherwise unknown to them and feel that their time was not wasted reading this book.

This book is divided into 11 chapters. The first chapter covers the early stages of warfare over the radio spectrum not involving radar. It can be considered the precursor of electronic warfare. The fundamental concepts of radar and some history about its early development are presented in Chapter 2. An overview of electronic warfare and an introduction to an EW system are provided in Chapters 3 and 4, respectively. Mr. Joseph Caschera, a retired US Air Force Research Laboratory (AFRL) engineer, provided valuable information for sections on EW processors in Chapter 4. Chapter 5 is dedicated to the countermeasures and counter-countermeasures taken by EW systems and radars. The main purpose of an EW system covered in this book is to prevent an aircraft being locked by hostile radars that guide missiles. Chapter 6 focuses on the detection of and defense against missiles after they are launched. Chaff, flare, and decoys can be considered as passive EW countermeasures as they do not emit jamming signals. They are introduced in Chapter 7. An EW system is designed to conduct countermeasures for disturbing radar operation. If an aircraft is invisible to radars then there is no need for such countermeasures as, in this scenario, radars present no threat, and stealth aircraft covered in Chapter 8 were developed for this purpose. Similarly, a low probability of intercept (LPI) radar was conceived to evade being detected by EW systems; thus, its operation will not be interrupted. Chapter 9 focuses on LPI radars. Machine learning has become a tool that people from every tech or non-tech field would like to play with, and Chapter 10 outlines some applications of machine learning in the field of EW. Chapter 11 concludes this book.

Both authors are greatly indebted to their colleagues and collaborators at the US AFRL which provided them unique opportunities to develop their expertise in the field of EW. It is impossible to list the names of all of AFRL colleagues who helped us to grow in our careers, but we deeply appreciate their support. The majority of this book was finished during Chi-Hao Cheng's sabbatical and he would like to express his gratitude for the backing of Miami University. Miss Shannon Cheng helped to illustrate this book and her efforts are sincerely appreciated. We also would like to acknowledge the assistance received from Ms. Nicki Dennis and Ms. Junko Nakajima of River Publishers. This book would never become a reality without their help. Last, but not least, we wish to thank our spouses, both of whose names are Susan, for their encouragement and understanding of our spending serious time on this book.

Notes from Chi-Hao Cheng

This book was initiated by a project Dr. James Tsui and I started in 2017. The original plan was to write a "story book" about radars and EW systems and some stories would be from Dr. Tsui's personal experiences through his long career with the US Air Force. Although Dr. Tsui invited me to join hands with him, he generously insisted on being the second author. Later on, we decided to add more technical content and finished several chapters. However, when we contacted potential publishers, the feedback we received was that we should instead work on a textbook on advanced EW systems for practicing engineers. Therefore, we postponed our project. Unfortunately, we were never able to start another book project as Dr. Tsui's health rapidly deteriorated and he passed away in 2019. Afterwards, I determined to expand and complete our unfinished project and Ms. Nicki Dennis kindly suggested that I publish this work with River Publishers. I hope this book lives up to her expectation. Dr. Tsui usually dedicated his books to his parents and/or wife, but this book should be a memorial to him. I am deeply humbled and greatly honored to co-author the final book of Dr. Tsui who brought me to the field of electronic warfare 15 years ago.

List of Figures

List of Abbreviations

ARH	Active Radar Homing
AM	Amplitude Modulation
AFRL	Air Force Research Laboratory
AWT&T	American Wireless Telephone and Telegraph Co
AAA	Anti-Aircraft Artillery
BPSK	Binary Phase Shifted Keying
CM	Carry Memory
CWD	Choi–Williams Distribution
CW	Continuous-Wave
CNN	Convolutional Neural Network
COSRO	Conical Scan on Receive Only
CRS	Cross Section
dB	Decibel
DRFM	Digital Radio Frequency Memory
DIRCM	Directional Infrared Countermeasures
EA	Electronic Attack
ECCM	Electronic Counter-Countermeasure
ECM	Electronic Countermeasures
ELINT	ELectronic INTelligence
EP	Electronic Protection
ESM	Electronic Support Measures
EW	Electronic Warfare
FM	Frequency Modulation
FMCW	Frequency Modulated Continuous Wave
FSK	Frequency Shift Keying
HARM	High-Speed Anti-Radiation Missile
IR	Infrared
IF	Intermediate Frequency
J/S	Jamming to Signal
LFM	Linear FM
LSTM	Long Short-Term Memory

LORO	Lobe on Receive Only
LPI	Low Probability of Intercept
MANPADS	Man-Portable Air Defense Systems
MALD	Miniature Air-Launched Decoys
PLAAF	People's Liberation Army Air Force
PRF	Pulse Repetition Frequency
PRI	Pulse Repetition Interval
PW	Pulse Width
QPSK	Quadrature Phase Shifting Keying
Q-Value	Quality Value
RAM	Radar-Absorbent Material
RF	Radio Frequency
RNN	Recurrent Neural Network
ROCAF	Republic of China Air Force
RAF	Royal Air Force
SARH	Semi-Active Radar Homing
Superhet	Superheterodyne
SAM	Surface to Air Missiles
TALD	Tactical Air Launched Decoy
TOA	Time of Arrival
TR	Transmit/Receive
TWT	Traveling Wave Tube
UV	Ultra-Violet
VHF	Very High Frequency
VOA	Voice of America
WVD	Wigner Ville Distribution

1

Introduction: From the First Jamming to the Battle of Beams

1.1 The First Jamming [1–6]

In 1901, the 11th America's Cup international yacht race was held in New York City. The race was postponed for several weeks due to the assassination of President William McKinley in September. That year, the race was between the defender, New York Yacht Club's *Columbia* and the challenger, Royal Ulster Yacht Club's *Shamrock II*. Two companies, Guglielmo Marconi's Marconi Company and Lee de Forest's Wireless Telegraph Company, placed wireless transmitters on boats hired for this event for transmitting bulletins to reporters on shore. Marconi was contracted by Associated Press and Lee de Forest worked for the Publishers' Press Association. In 1899, Marconi successfully covered the 10th America's Cup with his wireless equipment, and this success significantly boosted his company's value [1]. This time, de Forest aimed to achieve similar publicity by competing against Marconi directly.

As these two companies started to transmit signals, their messages were corrupted by a third company, American Wireless Telephone and Telegraph Co (AWT&T). Without securing any contract with a news agency, according to one account, AWT&T set up a scheme to jam both Marconi and de Forest's transmitters while transmitting its own updates on the yacht race [2, 3]. Equipped with a transmitter more powerful than ones used by its competitors, an AWT&T engineer named John Pickard used a simple code to provide real-time updates on the race, one repeated 10-second dash (a tone of long duration) for *Columbia's* taking the lead, two repeated 10-second dashes for *Shamrock II's* taking the lead, and three repeated 10-second dashes for neck-to-neck, etc. During these long dashes, no other transmitters' signals could be received correctly. As a result, neither Marconi nor de Forest's company could provide accurate reports on the yacht race. Nevertheless, AWT&T also failed to cover the yacht race (by the way, *Columbia* won the race). Some people consider this event to be the first deliberate jamming in the history of radio communications. However, based on other accounts, the interference caused by AWT&T was not intentional and the "jamming" was not successful. In this version of story, the first deliberate (and successful) jamming happened in the 1903's America's Cup with three identical players, the Marconi, de Forest, and International Wireless Telegraph & Telephone Company, the successor of AWT&T [4, 5]. In this version of story, the third player, Wireless Telegraph & Telephone Company, remains the villain.

In 1887, the German physicist Heinrich Rudolf Hertz demonstrated the existence of electromagnetic waves which was predicted by the Scottish scientist James Clerk Maxwell. James Clerk Maxwell's famous Maxwell equations provide a framework to describe electrical and magnetic fields and their interactions. Maxwell predicted the existence of waves of oscillating electric and magnetic fields propagating at the speed of light and claimed that light and electromagnetic waves are of the same nature. Using the setting shown in Figure 1.1, Hertz demonstrated that when a high voltage is applied through the coil, a radio wave can be generated and then received by the receiver illustrated in the same figure.

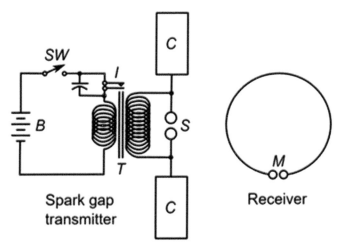

Figure 1.1   The circuit diagram of Hertz's experiment (https://upload.wikimedia.org/wikip edia/commons/8/83/Hertz_transmitter_and_receiver_-_English.svg).

Soon after the existence of electromagnetic waves was confirmed, people started to develop applications based on this discovery. For some engineers, electromagnetic waves can be a useful means for communications. Wire telegraphy that uses an electrical field through a wire to transmit information was invented in 1830s. Many inventors developed wire telegraphy independently, and one of them is an American inventor named Samuel Morse. Samuel Morse conceived the famous Morse code that uses combinations of tones of short duration (known as dots), tones of long duration (known as dashes), and a period of silence to represent the Latin alphabet and Arabic numbers (the length of silence depends on whether it is used to separate dash/dot, words, or a sentence). If an electrical field through a wire can be used to transmit information, it was only natural for people to try to develop a wireless telegraphy using electromagnetic waves through the air. Hertz's demonstration of electromagnetic waves provided great insights for wireless telegraphy developers, and the one who first showed feasibility of such a system is an Italian inventor, Guglielmo Marconi. Marconi first demonstrated that he could ring a bell remotely across a room using radio waves; he then continuously worked to extend the transmission distance. Marconi first demonstrated his radio telegraphy to the British government in 1896 and was awarded the Nobel Prize in Physics jointly with Karl Ferdinand Braun in 1909 "in recognition of their contributions to the development of wireless telegraphy." Among all of the prominent electrical engineers in his era including Thomas Edison and Nikola Tesla, Marconi was the only one awarded this prestigious award.

Since the invention of radio communications, its value was soon recognized by military. Shortly after Marconi demonstrated the first trans-Atlantic wireless communications in 1901, radio communications became a common practice in Navies around the world. Jamming one's enemy's radio communications was soon practiced in military exercises. On February 8, 1904, the Russo-Japanese War broke out and Port Arthur (now called Lüshun Port) was under siege in May 1904. During the siege of Port Arthur, Russian radio operators became aware of the increasing intensity of radio communications before Japan's attacks and used it as an early warning sign. On March 8, 1904, two Japanese armored cruisers *Kasuga* and *Nisshin* bombarded Port Arthur. A small Japanese destroyer close to the coast served as a forward observer to spot the location of where shells fell and radio gunfire corrections back to the two cruisers. Noticing this radio communication and understanding its importance, a Russian radio operator jammed the communications. Due to this effort, the damage caused by the bombardment was significantly reduced.

This event is generally seen as the first application of radio jamming in a real battle [5].

Because of the heavy loss suffered by Russian's Far East Fleet, Russian's Baltic Fleet was sent for enforcement. Since the Baltic Fleet has fewer ships and they are slower than Japanese Combined Fleet's ships, the priority of Baltic Fleet was to reach Russian Port of Vladivostok (Port Arthur fell on January 2, 1905) without being detected. On May 27, 1905, while approaching Tsushima Strait, the Baltic Fleet was spotted by a Japanese armed merchant cruiser Shinano Maru. The Shinano Maru reported this news to Japanese Admiral Togo Heihachiro's flagship using radio. Interestingly enough, although the Baltic Fleet was equipped with radio transmitters capable of jamming Japanese's radio communications and radio jamming has been practiced by Russians in the early stage of the war, Russian Admiral Zinovy Rozhestvensky decided not to jam Japanese's radio against his officers' advice. As a result, the Combined Fleet was able to locate Baltic Fleet and won a decisive victory in the Battle of Tsushima.

1.2 Radio for Navigation and Battle of the Beams [6–8]

Besides communications, radio signals can also be used as a navigation tool. Naturally, radio navigation was soon used in the war, and, not surprisingly, countermeasures and counter-countermeasures were developed. The earliest example is the so-called "Battle of the Beams" in the World War II.

After World War II broke out, Nazi Germany soon conquered most parts of Western Europe. Recognizing the difficulty of invading Britain from the sea due to Great Britain's mighty sea power, Nazi Germany planned to force Britain into a settlement for peace through a combination of sea blockades and air raids. Luftwaffe bombers were sent to destroy British military and economic targets including ports, air field, industry, etc. Compared with a fighter, a bomber is larger and heavier, thus being an easy target for a fighter. When conducting bombing, Luftwaffe bombers would need the escort of fighters to protect them from the attack of Royal Air Force (RAF) fighters. However, since the Luftwaffe was unable to gain air superiority over RAF in the Battle of Britain, its bombing needed to be conducted at night time. During the night, the RAF fighter's could not see the Luftwaffe bombers; so they were safe from attack. On the other hand, the Luftwaffe bombers had difficulty locating their targets in the dark. In order to find a certain target at night, the bomber must know its direction and distance from the target in the dark, and some non-visual navigation method was necessary.

An easy way to help a pilot find the correct direction is to use a beacon radio signal as a guide. The pilot can fly an airplane following the direction pointed by the radio signal like a person walks along a path illuminated by a flash light. Nevertheless, there is a technical issue in this approach. If the radio beam is narrow and a pilot is off the path, they will have huge difficulty in correcting their direction. If the radio beam is broad, since the beam is always divergent from its emitter, the directional accuracy will suffer.

To solve this issue, a two radio beam navigation system called the Lorenz system was developed before the World War II. The Lorenz system uses two proximate radio transmitters to transmit radio signals, one transmitting dash and one transmitting dot as shown in Figure 1.2. These two radio transmitters are synchronized so that the dot is transmitted in the silent period between dash and vice versa. These two radio beams also slightly overlap with each other; hence, when pilots fly in the overlapping area pointed from transmitters, they will hear a continuous tone. This overlapping area is called the "equi-signal area." If pilots ever wander off the path, they will know how to make a correction based on the radio signals they hear. This method was used to guide airplanes to land at night before World War II and most of the airplanes were already equipped with a receiver to receive a Lorenz signal of 30 MHz. To adopt this method to guide a Luftwaffe bomber flying toward Britain, more powerful radio transmitters which can generate a radio beam with narrow divergence were necessary because the Lorenz system is designed to guide airplanes flying toward the transmitter, but this military application required bombers to fly away from the transmitter and toward their target. Nevertheless, Germany successfully accomplished this task and built several such transmitters in occupied France, emitting radio signal toward Britain. The radio beam used for direction is referred to as

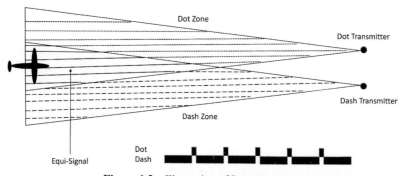

Figure 1.2 Illustration of Lorenz system.

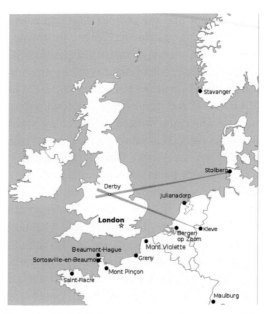

Figure 1.3   Map illustrating Knickebein system (https://commons.wikimedia.org/wiki/File: Map_of_Knickebein_transmitters.svg).

the approach beam. Besides the right direction, pilots needed to know where to drop bombs. Therefore, a second radio beam referred to as crossbeam was used. Germany set up another set of transmitters emitting crossbeams intersecting with the approach beam at the target as shown in Figure 1.3, and the bombers would drop bombs at the intersecting area. The Germans named this system Knickebein.

The British suspected that the Germans were using a radio navigation system, and Dr. R. V. Jones, a scientist working at Directorate of the Intelligence at the Air Ministry, guessed that Germany was using a navigation system similar to the Lorenz system. So, he ordered an airplane to search for any radio signals around 30 MHz. Some British scientists were very skeptical about the feasibility of such a long range radio navigation beam with 30 MHz frequency, as they did not believe that a radio wave of 30 MHz would bend over the curvature of the earth. But, after several failed searches, the German radio approach beam was found. Once the signals were identified, countermeasures needed to be taken. One solution was to jam the German beacon radio signals with strong noise. However, such a practice would alert the Germans that their system had been jammed. So, another countermeasure was taken. The British military designed a transmitter called "Aspirin" (the nickname the British military gave to Knickebein was "headache") which transmits a dash signal similar to the one transmitted by the Knickebein system with a directional antenna. In principle, if aiming the jamming signals into the Knickebein's dot zone, the British military can either create another equi-signal area of their choice or make German pilots think they were in the dash zone, thus misleading the German bombers. Also, if German pilots heard two dash signals, they would disregard the authentic dash tone for the dash tone generated by the Aspirin had a higher power. Because German pilots were trained to follow the approach beam and drop bombs at the spot where the approach beam and crossbeam intersect, when the Knickebein was jammed, many German bombers missed their targets by a large distance. Some British believed that their engineers were able to direct the German bombers to any direction and drop bombs at a place of their choice, so that when German bombs were dropped at Windsor castle, a complaint was made to the British government by the Comptroller of George VI's household.

The German Knickebein system was developed based on the Lorenz system used for civilian aviation before World War II. Both the Knickebein and Lorenz systems use radio waves of 30 MHz. A similar system named X-Gerät was also developed. The working principle of the X-Gerät system's approach beam is the same as the Knickebein, but the X-Gerät system uses radio signals of 74 MHz and its "equi-signal area" is much smaller than the Knickebein system, thus enabling a more precise bombing. The narrow approach beam of X-Gerät was code-named Weser.

The major difference between the Knickebein and X-Gerät systems was that, instead of one, the X-Gerät system used three crossbeams that intersected the approach beam at different locations as shown in Figure 1.4. When the bomber is 30 miles away from its target, the bomb aimer would hear a sound generated by the first crossbeam (code-named Rhine) and know that they were close to the target. Once the bomber passed the second crossbeam (code-named Oder) which crossed the approach beam 12 miles away from the target, the bomb aimer would hear another warning signal and then started a special stopwatch. The last crossbeam (code-named Elbe) intersected the approach beam 3 miles away from the target where the bomb aimer pressed the same button again. Based on the time elapsing between crossing the second and third crossbeams, the speed of the aircraft could be determined. Hence, if it took the bomber 30 seconds to fly across the 9 miles between the second and third crossbeams, the bombs would be dropped 10 seconds automatically after the button is pressed for the second time.

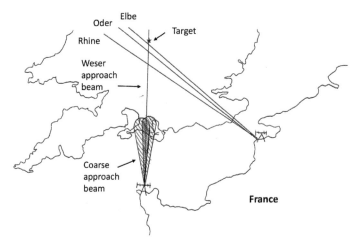

Figure 1.4   Map illustrating X-Gerät system (adapted from a map in https://www.oocities.o rg/pentagon/2833/general/tactics/xgeraet/xgeraet.html).

In principle, the X-Gerät system's approach beam can be jammed in the same fashion of jamming the Knickebein system. However, the rate of British jammer's dash signal is 1500 Hz and the rate of X-Gerät system's dash signal is 2000 Hz. As a result, the X-Gerät system's filter could filter out the British jamming signal. Eventually, the British military generated a false crossbeam to be detected by German bombers soon after the bomber crossed the second crossbeam. Therefore, the German bombers dropped their bombs before they reached their target. Since the X-Gerät system uses a radio frequency different from the Lorenz system's frequency, not every German bomber was equipped the X-Gerät system. In each mission, a few German bombers equipped with the X-Gerät system served as pathfinders and were followed by other German bombers. As a result, if the pathfinder was deceived, the bombing accuracy of the whole group would be compromised.

After the Knickebein and X-Gerät systems were jammed by the British, another German navigation system, Y-Gerät, was developed. The Y-Gerät system's ground station sent an approach beam aimed at the target for the bomber to follow, just like the previous Knickebein and X-Gerät systems. However, instead of transmitting crossbeams from other ground stations, the same ground station also sent a separate ranging signal to be picked up by the bomber. After receiving the ranging signal, the bomber would immediately retransmit the signal back to the ground station. By comparing the phase of the received and transmitted signals, the ground station could accurately determine the location of the bomber and sent a bomb release signal when the bomber arrived at its target.

The British jammed the Y-Gerät system simply by picking up the range signal sent by the bomber and retransmitted it to the German ground station so that the ground station could not determine the bomber's location precisely. After the Y-Gerät system was successfully jammed, the Germans began to move their bombers to Eastern Europe for the imminent attack on the Soviet Union and the Battle of Beams was over.

From this piece of early electronic warfare (EW) history, one can grasp some important EW principles. The first is that, unless necessary, do not transmit, as the adversary always searches for signals in the battle field. The Germans turned on their systems long before their bombing missions or even when they were not in use. If they had not transmitted signals when not being used, it would have taken the British longer to find the German guidance signals (some British experts did not even believe the existence of such signals in the beginning). Also, any knowledge about an adversary's signal is crucial. Dr. R. V. Jones made a correct guess about the German's navigation system signal frequency and looked for it. Any failed search for German radio signals would have put Dr. Jones in a difficult situation as he ordered search flights for German navigation signals despite objections from some of his colleagues. Developing countermeasures based on none or very little knowledge about interested signals' characteristics would be a daunting and time-consuming task, and in any kind of competition, very often, speed to respond determines the final outcome of the competition. Moreover, when jamming is conducted, the best practice is to mislead the enemy without causing any suspicion about being jammed. This might be the reason why the British did not simply use noise to jam the German navigation signals. Finally, one should consider how to counter the jamming signals. This area is generally referred to as electronic counter-countermeasures (ECCM). In the Battle of Beams, the Germans never tried to take any ECCM against British jammers.

1.3 The Scope of this Book

Wireless communications and radar are two early applications of electromagnetic waves and still play significant roles in today's world. Like some other technologies such as airplanes, wireless communications and radar were adopted for military use almost from their very beginning. The focus of this book is about radar and its countermeasures, EW system in air warfare.

The purpose of radar is to detect objects from a long distance, and EW technology is designed to detect the existence of radar and take appropriate actions to interfere radar signals. If a radar concludes that its signal has been compromised, a counter-countermeasure might be taken. Jamming examples described earlier in this chapter demonstrate how quickly and creatively people conceive a corresponding countermeasure after a tool is developed to gain commercial or military advantage. Although examples presented in this chapter are not about radar, they are often considered precursors of EW.

There is a 16th century Chinese novel, *Investiture of the Gods* (封神榜), which is about the war between the Shang dynasty and the Zhou dynasty of China happening around 1100 BC. In this novel, each side employed gods and demons in their fighting and many of them have special powers such as being able to listen to enemy's conversations and observe enemy's maneuvers from 1000 miles away. The demon with the special hearing power is called Qianliyan (千里眼) and the demon with much better than 20–20 vision is called Shunfenger (順風耳). Usually, one special power will meet its countermeasure just like radar and EW. In this story, Qianliyan and Shunfenger were employed by the Shang dynasty in the war and caused huge damages to Zhou's army. Initially, Zhou's field marshal could not find out how his battle plans and moves were leaked to his enemy until his intelligence source revealed the special powers of these two demons. Once Zhou's field marshal learned about these two special powers, countermeasures were then conceived: a group of 1000 drummers played drums to deafen Shunfenger and 2000 of red flags were used to block the view of Qianliyan during the preparation of battle. Thus, these two special powers were neutralized and Qianliyan and Shunfenger were eliminated in the next battle. Interestingly, the techniques described in this 16th century novel share some similar characteristics with EW techniques. Deafening the Shunfenger using drummers is similar to jamming a radar with strong noise so the radar cannot receive reflected radar signals and we can almost consider blocking Qianliyan with 2000 red flags a precursor of chaff which is used to induce plenty of false echoes on a radar screen thus blocking the radar's "view."

Through this book, if available, historical events are used to demonstrate radar and EW technologies. However, the aim of this book is to introduce fundamental radar and EW working principles rather than serve as a history of radar or EW. Besides some knowledge of high school or college freshman level mathematics and physics, it does not require an advanced technical background to comprehend the material in this book. The theme of this book is the measure and countermeasure between radars and EW systems.

We believe the story about the development of radar and EW technologies is as interesting as a good novel like *Investiture of the Gods* and understanding the principles of these technologies will bring about a deeper appreciation of human creativity. We hope readers will agree with us after finishing the book.

References

[1] Orrin E. Dunlap, *Marconi: the Man and His Wireless*, the MacMillan Company, 1937.

[2] Alfred Price, *The History of US Electronic Warfare*, vol. I, The Association of Old Crows, 1984.

[3] R. Schroer, "Electronic Warfare. [A century of powered flight: 1903–2003]," IEEE Aerospace and Electronic Systems Magazine, vol. 18, no. 7, pp. 49–54, July 2003.

[4] Thomas H. White, "Pioneering U.S. Radio Activities (1897–1917)," United Sates Early Radio History, http://earlyradiohistory.us/sec007.htm.

[5] Adrian M. Peterson, "World's First Jamming Transmissions" Wavescan, October 7, 2012, http://www.ontheshortwaves.com/Wavescan/wavescan121007.html.

[6] Mario de Arcangelis, *Electronic Warfare: From the Battle of Tsushima to the Falklands and Lebanon Conflicts*, Blandford Press, 1985.

[7] Alfred Price, *Instruments of Darkness: the History of Electronic Warfare 1939–1945*, Frontline Books, 2017.

[8] Carlo Kopp, "Battle of the Beams," Defence Today, pp. 76–77, January/February 2007.

[9] R. V. Jones, *Most Secret War*, Wordsworth, 1978.

2

Radar Fundamentals and Their Early Development

2.1 Introduction

RAdio Detection And Ranging (RADAR) was invented for military operations to detect an object's position and speed from a long distance. Today, there are many types of radar for civilian applications. For example, weather radar monitors weather condition using electromagnetic waves reflected from clouds. In airports, air traffic control radars are used to detect airplanes in the sky for air traffic monitoring/control. Police use radar speed guns to check drivers' speed. Although people might not necessarily like radar speed guns, they keep driving safe to some extent. There are also numerous specially designed radars for different applications such as ground-penetrating radar for mapping underground features like tree root systems, etc.

Electronic warfare was conceived as a countermeasure to radar for military use. Before the invention and deployment of military radars, electromagnetic waves had been used for communications and navigation for military purpose as described in Chapter 1 along with corresponding jamming practices. Nevertheless, these jamming practices should be considered as precursors to electronic warfare, and electronic warfare ought not to be considered as an independent area of research. Electronic warfare is a response to the application of radar in military, and only when radar was invented and its applications spread over all military applications, did countermeasures against radar (i.e. electronic warfare) became an important issue and research in electronic warfare started to thrive.

The Chinese phrase for contradiction is pronounced as mao-dun (矛盾). Mao (矛) means spear and dun (盾) means shield. There is a story written by a Chinese philosopher Han Fei (280–233 BC). It says a merchant was selling both spears and shields in a market. He claimed that his spear was so sharp that it could punch through anything and nothing can stop it. Then he took out

13

his shield and claimed that it can stop any weapon. A bystander then asked him what would happen if someone uses his spear to attack his shield and the merchant could not answer.

Most likely, Han Fei made up this story to explain the concept of self-contradicting statement. It is not uncommon in today's defense industry that one company makes both radars and electronic warfare systems but probably none of them has or will ever make such a bold statement as the one made by this ancient Chinese merchant in any military equipment exhibition. Nevertheless, some lesson can still be learned from this story. A spear maker's goal, of course, is to make the spear sharp enough to punch through any shield, and, to achieve this goal, some understanding about the shield is necessary. The same principle also applies to the shield makers. For the same reason, to understand electronic warfare techniques, knowledge about radar is necessary.

In this chapter, the fundamental concepts and a brief history of radar will be introduced. For a radar to become a reality from a concept, numerous difficult engineering problems must be solved first. A few of these key issues will be addressed in this chapter. Since this book is about electronic warfare, only military radar will be discussed.

2.2 The Difficulty of Estimating Distance

Technically, it is difficult to measure the distance of an object far away. A person might estimate his/her distance from an object via visually checking the size of object. If a person looks small, this person should be distant. Sometimes, the distance is estimated through the background such as saying "a person is about two blocks away." However, such a visual check is not always accurate. There is a Chinese saying: "Riding a horse toward a mountain can tire the horse to death" meaning that a horse can be exhausted by its rider for keeping it running toward a mountain while the distance to mountain is much further than it appears. Some cross-country runners might agree with this statement.

Chapter 1 describes some techniques developed by the German military to estimate flying distance using radio beams for night air raids during the Battle of Britain. Although the principles of these methods are correct, the accuracy cannot be great since the widths of radio beams used to indicate distance are too broad.

The invention of radar solves at least two important issues. One is to detect a distant target. Before the invention of radar, soldiers used binoculars to search for aircraft in the sky. If one cannot hear the engine sound, it is a difficult job. Some aircraft have light bulbs on their wings. The light can decrease the probability of the airplane being detected because it mingles the aircraft with bright sunlight. The other issue radar solves is to determine the distance and speed of the aircraft. If one looks at an airplane flying directly toward him/her, the airplane can appear as a stationary object in the sky. Under this condition, one can estimate neither the aircraft's distance nor speed.

2.3 Bat and Radar

When James Tsui had just turned into a teenager, his family moved to Taipei, Taiwan. He lived in a house near the edge of the city. There were rice fields in the neighborhood and many bats lived in the same house he inhabited. His family lived under the ceiling and the bats lived about it. Since the Chinese considered bats creatures that bring good luck, the Tsui family and the bats co-existed pretty well. Later, when James Tsui was in college in the 1950s, he was told that people invented radar from the inspiration of bats. Bats fly in the dark and they cannot see in darkness. Each bat sends out an acoustic signal. When the signal encounters an object, the signal is reflected from the object. The bat receives the echo and uses the echo to guide its activity. Each bat uses this process to navigate through the darkness and capture insects. Engineers took this idea and replaced the acoustic signal with electromagnetic waves to build radars. Some online videos show that thousands of bats fly in darkness. These bats all appear to fly in random patterns without colliding with each other and they can catch their prey in midair while flying. The bats can only work in relatively short distances though. On the other hand, radar is designed to detect aircraft hundreds of kilometers away. However, if there are many fast-moving aircraft within a confined airspace, collision might be inevitable and ground radars might have difficulty separating them as well.

2.4 The Early Development of Radar [1, 2]

Using the reflected electromagnetic wave to detect objects is the fundamental principle of radar. In the 1930s, several countries including the United States, Great Britain, Germany, France, Soviet Union, Italy, the Netherlands, and Japan were developing radar based on similar principles. The acronym *RADAR* was coined by the United States Navy in 1940. Developing radar is not an easy task. The working principle of radar is easy to understand

and its purpose cannot be clearer; however, many technical issues did not have solutions until the late 1930s and early 1940s. One thing worthy of mention is that even after the radar was developed and became operative, not everyone trusted this new equipment. On December 7, 1941, when the newly deployed US SCR-270 radar at the Opana radar site in Hawaii detected a large formation of approaching Japanese bombers 132 miles away, this warning was dismissed [3]. As we know, the Japanese bombed Pearl Harbor later, and this is one of the biggest US military disasters. Although people can continuously speculate on what could have happened, if the radar warning signal had not been dismissed, it should be noted that due to the fact that the radar did detect incoming Japanese aircraft, the US radar project was very well funded during the remainder of World War II. It received more funds than the Manhattan project which, of course, created atomic bombs, ending World War II. People working on the US radar project during World War II used to say that "the atomic bomb may have ended the war, but radar won it."

    In the following sections, we will discuss the requirements of building radar and introduce some enabling technologies for radar.

2.5 Principle of Radars and the Radar Equation

Radars are designed to detect targets and determine their distance. The basic concept is to send out an electromagnetic signal and detect its reflection from the target. Although radar intends to detect targets in all directions instantaneously, it is difficult to achieve this goal. The problem can be explained as follows. Assuming, a radar sends out a signal to all directions at the same time, the electromagnetic signal from this source will spread out as a sphere. The surface of a sphere can be calculated as $4\pi r^2$, where r is the radius of the sphere and also the distance between the source and the target. As a result, the energy reaching the target is $1/4\pi r^2$ of transmitted energy times the cross-sectional area of the target. If r is large, only a very small portion of the transmitting energy reaches the target. Assume that the reflection from the target (also known as skin return) is not directional either and it is also spread in the form of a sphere. The density of energy reaching the radar is then $1/4\pi r^2$ of the reflected signal's energy. As a consequence, the total energy reflected back to the source is proportional to $1/(4\pi r^2)^2$. This is usually a very small portion of the transmitted signal energy. Even if a directional antenna is used (i.e. the signal is not sent out in all directions), the signal power is still inversely proportional to r^4. The relation between radar's transmitted signal power and received signal power can be expressed in a radar equation shown below

$$P_{rec} = P_t\, G\,\frac{\sigma}{4\pi r^2}\,\frac{\frac{G}{4\pi}\left(\frac{c}{f}\right)^2}{4\pi r^2} = P_t G^2\frac{\sigma}{(4\pi)^3 r^4}\left(\frac{c}{f}\right)^2 \qquad (2.1)$$

where P_{rec} is power of the received signal, P_t is the power of transmitted signal, σ is the object's cross section area (i.e. the area that reflects radar signals), G is the antenna gain, the term of $\frac{G}{4\pi}\left(\frac{c}{f}\right)^2$ is the receive antenna's affective area, c is the speed of light, f is signal frequency, and r is distance between radar and its object.

Both the terms $1/4\pi r^2$ and $(1/4\pi^2)^2$ are very important for electronic warfare applications. When designing a radar receiver, the characteristics of signal such as expected frequency range and signal duration are known. Thus, the radar receiver can filter out unwanted signals and be designed with high sensitivity, but the energy of its received signal is proportional to $(1/4\pi r^2)^2$ of the transmitted signal energy. The target aircraft also has a receiver on board and it is called an electronic warfare receiver which is used to detect unknown radar signals. Because the electronic warfare receiver on the target aircraft does not have information on the radar signals, it needs to cover a broad bandwidth so it does not miss potential hostile radar signals. As a result, compared with the radar receiver, the electronic warfare receiver's sensitivity is moderate at best. However, the energy of the radar signal that reaches the electronic warfare receiver is proportional to $1/4\pi r^2$ of the transmitted energy, which is much stronger than the energy of reflected radar signal received by the radar receiver. In addition, the electronic warfare receiver intercepts the radar signal before the radar receives the reflected radar signal. Figure 2.1 demonstrates this relation. The same reasoning can be applied to

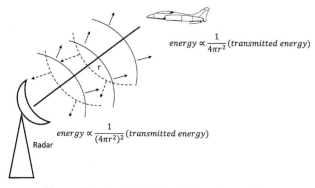

Figure 2.1   An illustration of the radar equation.

jamming. If the target aircraft wants to disturb the radar operation by sending a jamming signal toward the radar receiver, the energy of jamming signal reaching the radar is proportional to $1/4\pi r^2$ of the transmitted jamming signal energy. Therefore, the jammer can jam the radar using a signal much weaker than the radar signal. The subject of electronic warfare receivers will be discussed in Chapter 3.

2.6 Requirement of High Frequency Source

The concepts of wavelength and frequency of an electromagnetic wave can be explained as follows. Assume that the electromagnetic wave is a sinusoidal signal transmitting in the speed of light as shown in Figure 2.2. The wavelength is the length of one cycle of signal and the frequency is the number of cycles of signals passing a fixed point in 1 second. The relations between wavelength and frequency can be written as

$$f\lambda = c \tag{2.2}$$

where f is the signal frequency (unit: Hz), λ is the corresponding wavelength, and $c = 299,792,458$ m/s (meters per second) is the speed of light in a vacuum. As one can see from this equation, the higher the frequency, the shorter the wavelength. This can be explained intuitively as if the signal's propagation speed remains the same (in this case, the speed of light); the shorter the wavelength, more the cycles of the signal pass a fixed point within a given time frame.

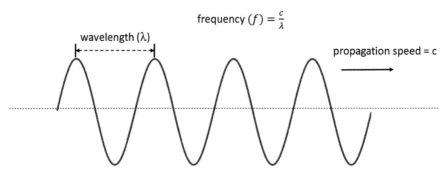

Figure 2.2   The relationship between speed, frequency, and wavelength of an electromagnetic wave.

In order to detect an object using an electromagnetic wave, the wavelength of the signal must be much smaller than the object. Thus, the signal wavelength cannot be very long. To detect the returned weak signal, one might increase the gain of the receiver antenna. As part of the antenna shape design, the antenna's physical size must be several wavelengths long. A movable antenna used in a battlefield cannot be too large. For example, if one limits the diameter of the radar target to about 1 m, the signal wavelength must be several times smaller than that and a wavelength of 1 m roughly corresponds to a frequency of 300 MHz. Therefore, the general military radar signal frequency is from 2 to 18 GHz and the corresponding wavelength is $0.15 - 0.0167$ m. Although a higher frequency can improve the resolution of the radar, some technical problems accompany signals of high frequency. First of all, the attenuation (loss) that the microwave signal suffers in air is rather low when the signal frequency is below 20 GHz. Above 20 GHz, the loss can be too high. By checking Equation (2.1), one can find that when the signal frequency increases, the received signal strength decreases in proportion to the square of the signal frequency. In addition, when the signal frequency increases, to maintain the same gain, the size of the antenna aperture (antenna's receiving cross section) needs to be decreased. When the frequency of a signal is doubled, the aperture needs to be decreased four times. As a result, signals with higher frequency will have a narrow signal beam width and it might be difficult to find the target (it is much easier to hit someone using a baseball bat than a long needle). In the 1930s, when radars were being developed, the generation of high frequency signal was a major technology challenge.

2.7 Requirement of Very Short Pulse

When a radar sends out a signal to detect objects, the signal duration (known as pulse width, PW) is finite. The radar sends out a signal first and then tries to capture the reflected signal before emitting the next radar signal. Based on the time the radar receives the reflected signal, the radar can determine the distance of this object. The resolution of radar's distance measurement depends on the radar signal's pulse width and this relation can be written as

$$\Delta d = \tau c / 2 \tag{2.3}$$

where Δd is the distance resolution, τ is the pulse width, and c is the speed of light. This equation can be explained as follows. A radar signal with a pulse width of τ will cover a distance of τC as the speed of an electromagnetic signal is the same as light. If the distance between two objects is shorter than $\tau C/2$, their reflected signals will overlap. Under this condition, the radar will have difficulty in separating them using the reflected signals. If the pulse width is 1 μs (microsecond) which is one millionth of a second, the distance resolution is about 150 m. This is not a very fine distance resolution. In addition, a radar does not use only one pulse to measure range. It uses many pulses. The separation between pulses depends on the maximum measurement range of the radar. If the maximum range is 150 km, the transmitted signal might travel twice this distance back to the receiver and that is 300 km. It takes about 1/1000 s for the transmitted signal to reach the farthest object within the radar's working range and be reflected back to the radar receiver. Therefore, the radar must wait for at least 1 ms (millisecond) before transmitting the next pulse. This time is called the pulse repetition interval (PRI) and its inverse is referred to as pulse repetition frequency (PRF). In this specific case, the value of PRF is 1 kHz. If the PRI is shorter than 1 ms, the radar might receive signal reflected by an object 150 km away after it sends out the second pulse. As a result, the radar might mistake this object as something very near. The received pulse train can be displayed on an oscilloscope. The ratio between PW and PRI is referred to as the duty cycle which means the ratio of time in which the radar transmission is on. For a radar system with a PW of 1 μs and PRI of 1 ms, its duty cycle is 0.001.

To generate a signal with a short pulse width such as 1 μs is not an easy task. Definitely, a mechanical switch is out of the question. In 1937, a British engineer, Alan Blumlein, invented a circuit to generate a short pulse and his design was later named the Blumlein transmission line [4, 5]. It uses propagation of electric charge through a transmission line to generate a short pulse. A conceptual Blumlein transmission line is illustrated in Figure 2.3. Initially, the switch is open, the voltage of whole transmission line is V_0, and the voltage across the load is zero. The load resistance (R_{load}) is twice the impedance of transmission line, l. When the switch is closed, the negative charge travels from the switch toward the load (R_{load}), half of the charge is reflected back to the switch, the other half propagates toward the open end, and the voltage across the load resistance becomes V_0. When the charge is reflected back from the open end toward the switch, the voltage across the load drops to zero. The time the charge takes to make a round trip from the load resistance to the open end is the pulse width. The velocity of current in the transmission line is about 60% of the speed of light in the air. To generate

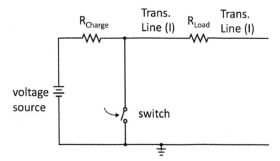

Figure 2.3   A diagram of the Blumlein transmission line.

a 1 μs pulse, the transmission distance is about 180 m. Since this method requires that the signal travels forward and back, the transmission line length is about 90 m. It should be clarified that, to generate a periodic short pulse, the switching frequency of this circuit does not need to be high as the PRF and is pretty modest (it is only 1 kHz in the previous example). This clever method solves the short pulse generation problem and similar circuits are still in use today.

Sections 2.6 and 2.7 describe a primary requirement of radar: generating a high frequency signal with short pulse width. To satisfy this requirement, special hardware needs to be developed.

2.8 High Power Microwave Source: Cavity Magnetron [6, 7]

As discussed in Section 2.5, a radar receives only a fraction of the transmitted energy. Energy equals signal power times the duration of the pulse. Since the pulse is very short, the microwave source must generate very high peak power, even at the megawatt level, although the average power is in the range of kilowatts. Cavity magnetron, a special vacuum tube, is used to generate high-power short-duration microwave radar signal. The working principle of the magnetron shown in Figure 2.4 is passing a stream of electrons through the openings of cavities in a magnetic field and generating an electromagnetic wave through interaction between electron and magnetic field. The frequency of the electromagnetic wave depends on the size of the cavities. In some sense, the way the cavity magnetron generates microwave signals of different frequency is similar to blowing air through the opening of a wind instrument to generate a certain note. An American engineer, Albert W. Hull, proposed the concept of using a magnetic field to control the electronic

Figure 2.4   Structure of magnetron. © 2018 Christian Wolff (https://www.radartutorial.eu/0 8.transmitters/Magnetron.en.html).

current and invented a magnetron tube around the 1920s. However, Hull's invention was not practical until two British engineers, John Randall and Henry Boot, significantly improved Hull's design in 1940 under the pressure of war. After World War II broke out, recognizing the importance of American assistance for Great Britain to win the war, British prime minister Winston Churchill decided to send a group of British envoys (known as the Tizard Mission) to the US, sharing British technology secrets including radar, jet engine, proximity fuse, and the idea of an atomic bomb, in September 1940 hoping the US could help perfect these technologies and assist Britain's war efforts. Not being able to generate high power electromagnetic signals was the major issue of the US radar project and it was not solved until the British government shared its cavity magnetron invention with the US. Randall and Boot's cavity magnetron could generate microwave signals with a power of 10 kW and wavelength of 10 cm (i.e. frequency: 3 GHz), while the US's radars at that point were operated at the wavelength of 1 or 2 m (i.e. frequency: 300 to 150 MHz). As explained in Section 2.6, a microwave signal with shorter wavelength can support higher resolution and reduce the size of the antenna. Learning how to generate high power and high frequency microwave signals from Britain was a key to the success of the US radar project in World War II. Readers interested about this part of history can refer to Jennet Conant's Tuxedo Park. The magnetron is a very important invention. Up to when this book was written (2020), it is still the most powerful microwave generator. Although solid state devices have been used to generate microwaves, the cavity magnetron still has the advantage in generating high power electromagnetic signals.

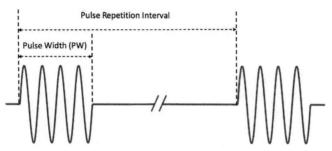

Figure 2.5 Pulse radar signal.

Once a high voltage short pulse is generated, this pulse can be applied to the cavity magnetron. The magnetron then generates a high power microwave signal. The output pulse train is shown in Figure 2.5. Each pulse is short and modulated by a high radio frequency (RF). The required peak power can be as high as many megawatts. This is an unimaginable value. The average microwave power of a microwave oven is 700 W. Even if the average power of a microwave oven is 1000 W, the pulse peak power equals the combined power of 1000 microwave ovens. However, if the pulse width and pulse repetition interval are taken into consideration, the average power of the radar is quite reasonable. For example, if the pulse peak power is 1 MW, the pulse width is 1 µs, and the pulse repetition interval is 1 ms, the average radar power is only 1 kW, that is, 1 MW divided by 1000 because the radar only transmits the signal for 1 µs over every 1 ms.

2.9 A Diversion: From Radar to Microwave Oven [8, 9]

Percy L. Spencer was a researcher at Raytheon in Waltham, MA, USA. He never finished elementary school, but, through self-learning, he became an expert in radio technology. One day, he worked in front of a radar and a peanut cluster in his pocked melted. This is not the first time such a thing happened, but he was the first one who decided to investigate. So, he put food such as an egg and corn kernel under a magnetron. The egg exploded and his face was covered with egg, but he was able to share popcorn with his officemates.

Everybody knew that the radar beam possesses a large amount of energy. People were even concerned that when a bird flew into the radar, it might get hurt. But, nobody thought that microwaves could be used for cooking. Even if somebody had thought of it, the approach was not practical because microwave energy generation is not trivial and is expensive. Spencer worked on the idea, and, in 1946, the microwave oven was invented. Of course, back then, the microwave oven was very expensive and had very limited applications.

Litton took over the microwave oven production in 1960 and spent a tremendous amount of effort developing it into a popular kitchen product. In the 1970s, it became a common kitchen appliance. A microwave oven is a cavity that microwaves oscillate in. Metals cannot be put in it because the metal will heat up and become red hot. James Tsui had personally experienced this effect. Once James Tsui made a fishing lure and pained it silver. In order to dry the lure quickly, James Tsui used a microwave oven to dry the hook and forgot the hook was made of metal. When James Tsui opened the oven, the hook was red and a very small hole was burned at the bottom surface of the oven. Fortunately, the microwave oven was not damaged. On the other hand, there is metal support in some microwave ovens. The microwave distribution in these ovens must be carefully analyzed so that the metal support will not be heated up. Nowadays, every modern kitchen has a microwave oven. Even many motel rooms are equipped with one. From its original invention to a popular commercial product, it took engineers over 30 years to prefect the microwave oven design.

Many inventions are originated from accidental discoveries such as penicillin and black rubber (adding carbon in rubber). A good researcher notices unusual phenomena, and, with some luck, useful products might be invented from the curiosity.

2.10 A Basic Radar System

A radar can be considered as consisting of four major parts. A microwave power source generates a pulse modulated microwave signal. An antenna is used to transmit the signal into air. The same antenna also receives the reflected signal from the target. The third part is a radar receiver to receive the signal from the antenna (A circulator is used to pass the received signal to the receiver rather than the transmitter.) and convert the microwave signal into video signals. The last part is the display unit. Usually, a cathode ray tube is used to display the transmitted and received signals. The time difference between the transmitted and received signals can be used to calculate the distance of the target. The basic radar schematic diagram is shown in Figure 2.6.

It should be emphasized that many details are omitted in this diagram. For example, the antenna might not be used to transmit and receive signals simultaneously. It might be time shared between transmitting and receiving. A transmit/receive switch (known as a TR switch) is used for this operation.

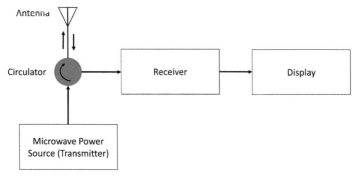

Figure 2.6 Basic radar schematic diagram.

2.11 Frequency Modulation Radar

Microwaves often travel in waveguide which is a rectangular-shaped metal pipe. Since when the microwave travels in air the loss is small, a high power microwave can cause the air to break down inside the waveguide, which is equivalent to a short circuit. To avoid this problem, different kinds of gas rather than air, or even liquid, can be used to fill the waveguide. Either way, this kind of system is difficult to maintain because keeping the gas or liquid from leaking through the waveguide is another problem to deal with.

An alternative solution to this problem is to reduce the peak transmitted microwave power and increase the pulse width to maintain the total energy. One way to achieve this goal is to use a frequency modulated (FM) signal. The frequency of an FM signal can start with a low frequency and end with a high frequency. At the receiver side, a device called a dispersive delay line is used to handle the FM signal. The delay time of dispersive delay line is frequency dependent. In this case, the delay time is long for the low frequency and short for the high frequency. If the FM signal passes through the line, the signal can be compressed into a short pulse because the low frequency part takes longer time and the high frequency part takes shorter time and they come out of the line at the same time. Therefore, the difficulty of generating and transmitting short pulse signals with high power can be overcome. The difference between starting and ending frequencies (known as bandwidth) and the pulse width can be used to calculate the processing gain. If the pulse width is 100 μs and the frequency bandwidth is 100 MHz, the product of these two quantities is $100 \times 10^6 \times 100 \times 10^{-6} = 10,000$, and this number is the processing gain. This quantity is also referred to as the time bandwidth product. Since a continuous wave (i.e. a microwave with a fixed frequency) has zero bandwidth, it has no processing gain.

This kind of radar must have a wide bandwidth to process the signal and that is the price one has to pay.

2.12 Searching Radar

As explained in Section 2.5, the skin return can be very weak compared to the transmitted signal, so the radar antenna must have a very high gain to increase the received signal strength. Also, as described in Section 2.10, the radar uses the same antenna to transmit and receive signals.

If the antenna is omni-directional, it can detect targets in all directions instantly. However, the gain of this antenna is 1 (or we can say that it has identical gain in every direction). An antenna gain is often represented in decibels (dB) and the decibel value is calculated as

$$dB = 10\log_{10}(Gain). \tag{2.4}$$

If the gain of an antenna is 1, we can also say its gain is 0 dB. When the antenna is omni-directional, if a target is detected, the direction of the target cannot be determined as the reflected signal can be from any direction.

To solve this ambiguity, instead of an omni-directional antenna, a directional antenna might be used. A radiation pattern of a directional antenna which has a main beam (main lobe) and side lobes is shown in Figure 2.7. The main beam of antenna is the area where transmitted/received signal receives maximum amplification. When a directional antenna is used in a radar system, the radar mainly looks for objects in the main beam. The side lobe is the area where signals are still transmitted or received, but the amplification signal receives is much weaker.

One might want to use a high gain directional antenna for a radar system. If a target is detected, the position of the target can be easily determined

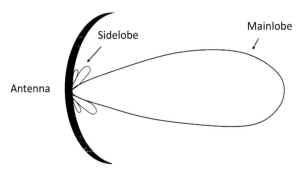

Figure 2.7   Radiation pattern of a directional antenna.

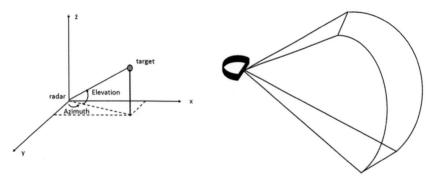

Figure 2.8   An antenna generating a vertical fan-shaped beam.

because the target is in the direction of the antenna's main beam. However, this arrangement can be difficult to use for detecting a target because the antenna gain and size of main beam size are related and higher gain means a narrower main beam. The chance of hitting a target using a microwave signal with a very narrow main beam is, of course, small.

The compromised approach is to use an antenna with medium gain and covering a reasonable angle. For example, a common approach is to create a vertical fan-shaped beam. The beam is narrow in the azimuth direction and wide in the elevation direction. To form this type of beam, the antenna will be wide in the horizontal direction but short in the elevation direction as shown in Figure 2.8. The beam shown in Figure 2.8 is called the fan beam and a beam that is narrow in both azimuth and elevation directions is referred to as the pencil beam. This antenna will rotate in horizontal direction. If a target is detected, the azimuth angle is known. The elevation can be found from a different antenna directed in the desired azimuth direction.

The returned signal is displayed on a cathode ray tube in polar coordinate. The location of the radar is at the center. A radiation line is rotating around the center. Detected targets are shown as bright points on the display. The target's azimuth angle and distance can be determined by its coordinate on the display. Figure 2.9 shows an example of the display.

2.13 Conical Scan Radar [10]

Once a target is found, its detailed information including target's position and velocity are needed. The type of radar gathering such information can be referred to as the tracking radar. The searching radar described in Section 2.12 can provide the position of a target. The target's velocity can be determined

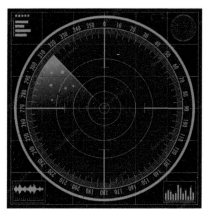

Figure 2.9   A radar screen. (https://static5.depositphotos.com/1000134/471/i/950/depositp hotos_4713955-stock-photo-radar-screen.jpg).

from the history of the reflected signals. For example, the target's velocity can be estimated via dividing the change in target position between two specific scans by the time interval between these two scans. Different approaches can be used to improve measurement accuracy, and some of them will be introduced in the following sections. First, let us discuss the conical scan radar.

The conical scan radar is used to find the accurate direction of the target and keep tracking it. Its antennas are usually horn-shaped. They are made of a rectangular waveguide and the dimensions of the waveguide increase into a rectangular horn, as shown in Figure 2.10. The conical scan radar's signal

Waveguide

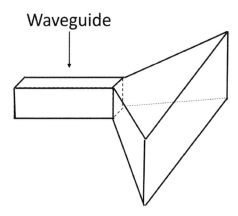

Figure 2.10   A rectangular horn antenna.

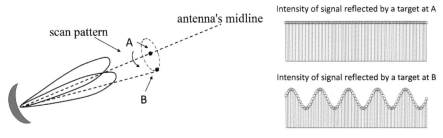

Figure 2.11   Conical scan illustration.

beam is slightly off center from its antenna's midline and continuously rotates around it (scan). If the target is roughly at the boresight of conical scan radar's antenna, the signal beam from the antenna makes a conical cone around the target, as shown in Figure 2.11. If the target is at the center of the cone, the reflected signal is strong and has constant magnitude during the rotation. If the received signal does not have constant amplitude during the rotation, the target is not at the center of the cone. An error signal is then generated to adjust the direction of the antenna. This approach can improve the accuracy of the direction measurement.

2.14 Monopulse Radar

This type of radar usually uses four horn antennas to transmit and receive four signals. In order to explain this idea, a system with two antennas is used for illustration. In this specific case, two antennas emit two signals. The two antennas point at slightly different directions. The reflected signals are summed together to obtain the sum channel. Their difference is also obtained as the difference channel. If the target is in the middle of these two beams, the sum channel will have the maximum output and the difference channel should have the minimum output. The outputs of the sum and difference channels are shown in Figure 2.12. The sum channel provides the coarse information about target direction and the difference channel provides the fine information. The output of the difference channel can be used to control the antenna's direction.

In the system of two antennas, the difference channel can only be used to adjust the two antennas in one direction such as in the azimuth direction. If four channels are used, the antennas can move in both the azimuth and elevation directions. Figure 2.13 shows the arrangement of four horn antennas. The channels A and C can be summed and B and D can be summed. Their output (A+C) and (B+D) can be considered as two channels to move antennas

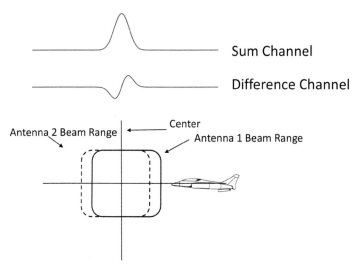

Figure 2.12 Monopulse radar with two antennas (for explanation purpose).

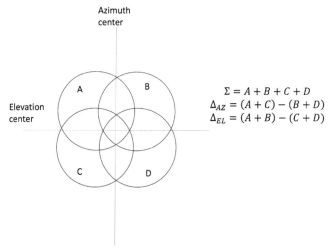

$$\Sigma = A + B + C + D$$
$$\Delta_{AZ} = (A + C) - (B + D)$$
$$\Delta_{EL} = (A + B) - (C + D)$$

Figure 2.13 Monopulse radar with four antennas.

in the azimuth direction. Similarly, the outputs (A+B) and (C+D) can be considered as two channels to move antennas in the elevation direction. Two difference channels, azimuth difference channel (Δ_{AZ}=(A+C)–(B+D)) and elevation difference channel (Δ_{EL}=(A+B)–(C+D)), can be used to precisely locate the target.

Compare the conical scan and monopulse radars; their major difference is that the former uses one antenna to find the maximum reflected signal and constant signal magnitude during rotation. The latter approach uses four antennas to find the maximum return signal and difference channels are used to guide antennas to track the object. Since the four antennas all work on a single pulse, this system is named a monopulse radar. The monopulse radar is more popular than the conical scan radar because it is less vulnerable to jamming. The topic of jamming will be discussed in Chapter 5.

2.15 Doppler Radar

Besides the target's location it is desirable to measure the speed of the target. Theoretically, the target speed can be calculated from the difference between the measured target locations within a certain time. The speed estimation obtained this way is coarse and not real-time (as the receiver needs time to acquire two distance measurements). Another way to measure an object's velocity is through the Doppler effect. The Doppler effect can be experienced in daily life. For example, one might hear a fire truck or police car with its siren on. One can hear a higher pitch when the truck or car is approaching and a lower pitch when it is moving away. This example demonstrates the Doppler effect of acoustic waves. The same principle also applies to electromagnetic waves.

Police radar uses the Doppler effect to measure the car speed and military radar uses the same principle to measure the aircraft speed. If the aircraft is approaching the radar, the received signal has a frequency higher than the transmitted signal. If the aircraft is moving away from the radar, the received signal has a frequency lower than the transmitted signal. By measuring the frequency difference between transmitted and received signals (also called the Doppler frequency), the speed of the target can be obtained as

$$s = \frac{\Delta f c}{2f} \tag{2.5}$$

where s is the target speed toward the radar, Δf is the difference between transmitted and received signals' frequencies, c is the speed of light, and f is the frequency of the transmitted signal. The factor of 2 in the equation is because the signal travels twice the distance between the target and radar.

If the target is traveling directly toward or away from the radar, the speed obtained with Equation (2.5) is the target's speed. If the target is traveling in another direction, the angle between the line of sight toward the radar and

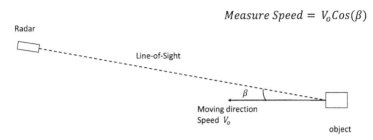

Figure 2.14   The effect of angle on the speed measured using Doppler effect.

the target's traveling direction must be taken into consideration. If the target travels in a tangential direction, the measured speed is zero, as explained in Figure 2.14. In other words, if there is a speeding vehicle with its side facing the police radar, the radar gun cannot measure the speed of the car.

One may think that the search radar can send RF pulse trains to measure the distance and use the Doppler frequency to measure speed of the target. There is a fundamental limitation in this approach though. According to Heisenberg's uncertainty principle, one cannot measure both the position and velocity (or momentum) accurately. In order to measure a distant target, the pulse separation must be long (i.e. the pulse repetition frequency is low) to eliminate range ambiguity, as discussed in Section 2.7. This design decreases radar's measurement ability in the frequency domain and the Doppler radar determines the target's speed based on the frequency shift caused by the Doppler effect. To improve the frequency measurement capability, the pulse separation must be short (i.e. the pulse repetition frequency is high). Thus, Doppler radar has high or medium pulse repetition frequency.

2.16 Continuous-wave Radar

Most of the radars considered in this book are pulse radar but continuous-wave (CW) radar is also used in some scenarios. The CW is refers to a continuous microwave signal with a fixed frequency. One might picture a CW signal as a sinusoidal wave. Unlike the pulse radar described in Section 2.7 which sends out a short pulse and waits for its skin return before sending the next short pulse, the CW radar continuously emits a CW signal. If the object is moving, the frequency of reflected signals will change, as explained in Section 2.15, and the CW radar can then detect the object based on the Doppler effect. The major advantage of a CW radar is its easy implementation since it does not have to generate a strong short pulse signal. Moreover, as a CW radar does not have to concentrate all of its energy in a short pulse, a CW radar can use a low power signal which is more difficult to intercept compared with the pulse radar's signal. Although the CW radar has difficulty detecting stationary or slow moving objects, and it cannot determine the distance of an object, it is very suitable for detecting fast moving objects such as a jet fighter. It also finds applications in speed guns.

2.17 Moving Target Indicator

Most of the time, radars point to the sky and any reflected signals are assumed to be reflected by targets. If radar is looking for low elevation targets and there is a high mountain in that direction, the reflection from the mountain may mask the true target. This unwanted reflected signal is called clutter. The clutter may be bigger than skin return from the true target because the mountain is much larger than the target. However, the mountain is stationary, so the reflection is at a fixed point. A delay line can be used to delay the received reflected signal. The amplitude and polarity of the delayed signal can be adjusted to cancel the returned signal reflected from stationary objects such as mountains.

Modern radars use coherent signal which means the phases of the transmitted signal can be controlled and the phase of received signal can be determined by the receiver. In this scenario, for a fixed target, the relation between the transmitted and received signal phases is a constant, but, for a moving target, this phase relation keeps changing. Comparing the phase relation between each transmitted and reflected signal, the radar screen can display only moving targets. Under a noisy condition, this approach has a better result than the amplitude cancelation approach.

2.18 Look-down/Shoot-down Radar

All the radars discussed in the previous sections are land- or sea-based radars looking toward the sky. The look-down/shoot-down radar is an airborne radar. It is a weapon radar and its goal is to look down toward the ground and find moving targets below the aircraft. Once the look-down radar identifies a moving target, it may then shoot at it (shoot-down), as shown in Figure 2.15. Since the radar is moving, even a stationary target such as a mountain becomes a moving object. The concept of moving target indicator described in Section 2.17 can no longer be applied here. Instead, the look-down radar distinguishes the moving target and stationary objects by comparing the different Doppler shifts of radar signals reflected from moving and stationary

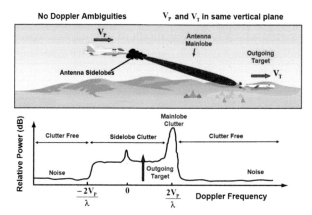

Figure 2.15   Doppler shift for look-down radar (https://docplayer.net/51967130-Radar-sys tems-engineering-lecture-14-airborne-pulse-doppler-radar.html).

objects [11]. As the speed of an aircraft is known, the Doppler shift of stationary objects can be easily determined and this information can be used to separate moving targets from stationary objects.

The ground ahead of the radar produces large reflection back to the radar. As a result, the clutter might mask the target. Since the ground contains various objects, e.g. buildings, hills, and plants, the clutter is not predictable. Signal processing must be used to reduce the clutter to increase the signal-to-noise ratio of signals reflected by targets and recognize them. This type of radar can be rather difficult to build.

2.19 Weapon Guidance Signal

In a battle field, the final goal of the radar is to find hostile targets and destroy them. Once a radar finds a target (let us assume that it is an aircraft) and obtains all the information about it, such as its location and speed, missiles might be launched toward the target to destroy it. The missile must be guided toward the target because the target might not move in a predictable pattern. In particular, when the target discovers that a missile is approaching, it will maneuver to avoid the attack and the missile must chase the target. The missile and the target aircraft have a tough game to play. There are many different ways the missile can be guided toward the target aircraft and they will be discussed in later parts of this book. One approach is to use a tracking radar, sending a guidance signal toward the target aircraft and the radar's signal will be reflected back to guide the missile. The missile then follows the reflected

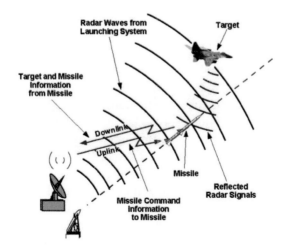

Figure 2.16   Semi-active radar homing (http://www.aerospaceweb.org/question/weapons/q0 187.shtml).

guidance signal toward the aircraft. This idea is shown in Figure 2.16. In the system illustrated in Figure 2.16, the missile detects the signal generated by an external radar and reflected by the target. Such a missile guidance method is referred to as the semi-active radar homing (SARH). Some missiles might have its own radar transceiver to send out guidance signals. This arrangement is referred to as the active radar homing (ARH).

2.20 Radar Modes

Three modes of radar considered in this book are: *search*, *acquisition*, and *tracking*. In the phase of search, a radar scans through the air, finding a potential target. Once an object is detected, in the acquisition phase, the radar scans vicinity of the detected object to determine the object's location and speed. After this information is obtained, the radar enters the tracking mode to actively track the object. It is worthy of note that modern radars are capable of engaging with multiple targets; so a radar might perform multiple tasks such as search and tracking simultaneously.

2.21 Conclusion

In this chapter, the principles of radar and basic technologies needed to build radar are discussed. Different types of military radar with different goals and operations are also presented. The principles of operations and limitations of the radars are addressed. However, all the discussions are on very fundamental operations. The radar performance depends on the development of the aircraft. For example, the development in stealth aircraft, which have small radar cross section, reduces the effectiveness of radars. Nevertheless, stealth aircraft do not absorb all the radar signals, although the research in this area has lasted for years. The radar and aircraft's radar cross sections are like the spear and shield problem mentioned at the beginning of this chapter. This is a problem without an end.

If the target of the radar, the aircraft, intercepts the radar signal, it will try to defeat the radar, and the actions the aircraft takes are called countermeasures. Then, the radar must take action to protect itself from hostile operations, and the radar's response is called counter-countermeasures. Usually, the signals generated by military radars are kept secret. Unless necessary, a military radar does not transmit. There are intelligence collection operations to collect military radar information even in peacetime. If one knows the adversary's military radar signals, the effect of the radar will be degraded. This point will be clearer when we discuss some real battle examples in later chapters.

Many books on electronic warfare focus on electronic warfare technologies designed to defeat or protect pulse radars. In recent years, low probability of intercept (LPI) radar, which is designed to avoid being detected by an electronic warfare system, has gained more popularity and some are a variant of CW radar or a pulse radar with long duty cycle. LPI radar will be covered in Chapter 9.

References

[1] Oskar Blumtritt, Hartmut Petzold, and William Aspray (ed.), *Tracking the History of Radar*, IEEE-Rutgers Center for the History of Electrical Engineering, 1994.

[2] Massimo Guarnieri, "The Early History of Radar," IEEE Industrial Electronics Magazine, vol. 36, no. 4. pp. 36–42. Sep. 2010.

[3] Patrick Hindle, Richard Mumford and Gary Lerude, "The Infamous Pearl Harbor Radar," Microwave Journal, May 12, 2017.

[4] Andrea de Angelis, Juergen Kolb, Luigi Zeni, and Karl H Schoenbach, "Kilovolt Blumlein Pulse Generator with Variable Pulse Duration and Polarity," the Review of Scientific Instruments, vol. 79. no. 4, April 2008

[5] S. J. Voeten, *Matching High Voltage Pulsed Power Technologies,* Technische Universiteit Eindhoven, 2013.

[6] Y. Blanchard, G. Galati, and P. van Genderen, "The Cavity Magnetron: Not Just a British Invention [Historical Corner]," IEEE Antennas and Propagation Magazine, vol. 55, no. 5, pp. 244–254, Oct. 2013.

[7] Samuel Y. Liao, *Microwave Devices and Circuits*, 3rd edition, Prentice-Hall, 1990.

[8] J. M. Osepchuk, "A History of Microwave Heating Applications," IEEE Transactions on Microwave Theory and Techniques, vol. 32, no. 9, pp. 1200–1224, Sep. 1984.

[9] Matt Blitz, "The Amazing True Story of How the Microwave Was Invented by Accident," Popular Mechanics, Feb. 24, 2016 (http://www.popularmechanics.com/technology/gadgets/a19567/how-the-microwave-was-invented-by-accident/).

[10] J. Litva, *Theory of Conical-Scan Radars for Low-Angle Tracking*, Defense Technical Information Center, 1980.

[11] W. H. Long, D. H. Mooney, and W. A. Skillman, "Pulse Doppler Radar," in *Radar Handbook*, M. I. Skolnik, ed. McGraw-Hill, 1990.

[12] Jennet Conant, *Tuxedo Park: A Wall Street Tycoon and the Secret Palace of Science That Changed the Course of World War II*, Simon & Schuster, 2003.

3

Overall View of Electronic Warfare

3.1 Introduction

Radar is a powerful piece of equipment for detecting aircraft. The days of looking for aircraft in the sky with the human eye are of course long over. In air warfare, in most scenarios, a jet fighter locates an adversary's jet fighter through an airborne radar at a relatively long distance and fires at it. The short-distance air dogfight between jet fighters portrayed in movies probably will happen less and less as those who are able to detect and lock the enemy at a longer distance might eliminate their enemy before their enemy sees them. Therefore, it can be safely stated that radar is a major threat to aircraft. It finds them, tracks them, and often guides missiles to destroy them. The radar can be considered as the eye of military operations. The main subject of this book is electronic warfare. The radar alone does not create electronic warfare, but it can be considered as the origin of electronic warfare, or the first layer of electronic warfare.

Electronic warfare operations can be divided into four layers, as illustrated in Figure 3.1. The first layer is radar searching and tracking operations discussed in Chapter 2. The second layer of electronic warfare is the receiver's radar signal interception and radar classification operation. After the radar signal is identified, the third layer of operation, countermeasure, might be used to neutralize hostile radar. The last layer of electronic warfare is radar's counter-countermeasure to protect itself. This first layer was covered in Chapter 2 and this chapter provides a brief introduction to the operation of layers 2–4. More details will be provided in the following chapters. People might name these operations differently and there are no universal accepted definitions. Different technical terms may refer to similar operations with slight difference and these different names may be only familiar to electronic warfare professionals. In this book, the names for the four layers of operations are simply the authors' preferences. If other technical terms are used in discussions, they will be explained.

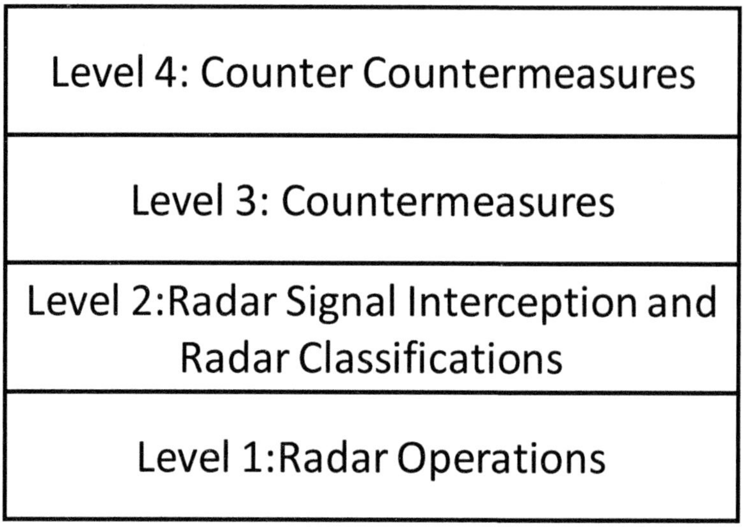

Level 4: Counter Countermeasures
Level 3: Countermeasures
Level 2:Radar Signal Interception and Radar Classifications
Level 1:Radar Operations

Figure 3.1   Four layers of electronic warfare operations.

Both the second and third layers of electronic warfare operations are performed on the target aircraft. The second layer of operations is called electronic support measures (ESM) by some researchers. They are also referred to as passive electronic warfare because these operations do not emit signals. Radar is an active piece of equipment because it transmits signals to find objects, thus making passive electronic warfare possible. It is worthy of mention that some passive radars that rely on signals of opportunities like TV and radio signals to detect objects without emitting signals exist, but, to the best of authors' knowledge, they are not capable of guiding missiles causing damage and thus are only discussed briefly in this book. The target aircraft must determine whether it is being illuminated by a hostile radar or not. If it is indeed illuminated by a hostile radar, hopefully, the radar cannot see it. As shown in Equation (2.1), the energy of signal reflected back to the radar is proportional to target's cross section area; so a radar can only detect targets with cross section large enough. It should be emphasized that a target's cross section area is not necessarily proportional to its physical size. Technology used to reduce the target's probability of being detected by radar, such as reducing the aircraft's cross section with a specially designed body shape or using special paints to absorb radar signals, are called stealth technology. Although the stealth technology is an important part of aircraft design, it cannot be ascribed to one of the four layers of electronic warfare operations as the stealth technology has not been developed to detect the existence of a radar signal or jam the radar. In other words, besides the four electronic warfare layers described here, there are related technologies such as stealth technology worthy of discussion and they will be discussed in Chapter 8.

The third layer of electronic warfare operation is to protect the target aircraft by interfering radar operations. The operations include sending noise to mask the target of the radar (jamming) or false signals to mislead the radar operators (spoofing). Since signals are transmitted in this operation, sometimes this is referred to as active electronic warfare. This operation might also be called electronic attack (EA). A more aggressive approach is to physically destroy the radar.

The fourth layer of operation is the radar's reactions to protect itself against hostile operations the target aircraft is conducting. Its purpose is to mitigate the effect of the interfering signals from the target aircraft.

3.2 Second Layer Operations: The Radar Signal Intercepts Performed on the Aircraft

An aircraft tracked by a radar not aware of the radar's operation could be in a great danger. Therefore, the operator of the radar's target aircraft must develop defensive techniques to protect itself. First, one must determine whether one's aircraft is being illuminated by a radar. A receiver is needed to receive all different radar signals and this type of receiver is referred to as the electronic warfare receiver or intercept receiver.

The main difference between an intercept receiver and the radar receiver discussed in Section 2.10 is that the radar receiver can distinguish the radar signal it is looking for, but the intercept receiver might assume no knowledge about radar signals. The radar receiver needs only to receive one signal, but an intercept receiver needs to receive all the radar signals in a battlefield. Worse yet, some radars are designed to avoid being detected by an intercept receiver and they are called low-probability-of-intercept (LPI) radar. Some LPI radar working principles will be discussed in Section 3.8 and Chapter 9.

Since the intercept receiver needs to cover a broad bandwidth, it receives noise/interference from a wide range of frequencies, thus having much lower sensitivity than a radar receiver. A receiver with higher sensitivity can detect weaker signals. However, the signal received by the intercept receiver is much stronger than the skin return as described in Section 2.5; so the intercept receiver with sensitivity lower than the radar receiver can still fulfill its designed goal. This is why some people might claim that electronic warfare is based on that single equation, Equation (2.1), in Section 2.5.

Although electronic warfare is a specialized field, people can actually find similar products in everyday life. The traffic radar detector used by some drivers to avoid speeding tickets is a kind of intercept receiver or a radar warning receiver. Its goal is to detect the existence of a police radar gun before the police can measure the vehicle's speed. Some people joke that a poorly designed traffic radar detector can only provide a driver extra time to reach his/her wallet because when the radar detector detects the radar, the police has already obtained the vehicle's speed.

3.3 Second Layer of Operations: Signal Classification and Identification Performed at the Aircraft

In a battlefield, there may be many different types of radar. Some are friendly and some are hostile. The intercept receiver is designed to receive all signals reaching the aircraft. If the intercept receiver is at high altitude, it will be illuminated by many types of radar. Sometimes, an intercept receiver can receive tens of thousands or even a million signals per second. However, only threat signals are of interest. The signal classification conducted by the electronic warfare processor begins with grouping all signals from a certain radar together. Once all signals from one radar are collected, the parameters of these signals, i.e. the frequency, the pulse width, and the time between pulses, can be extracted. These parameters are then used to identify the type of radar with the help of a radar database. If the radar is a threat radar, some actions must be taken. If the signal is guiding a missile toward the aircraft, the response time is in the order of seconds. On the other hand, if the radar is not an imminent threat, action may not be needed.

Of course, one must establish a library of radar signals and the library must have all the information on hostile radars. It is not an obvious job. The signal collection may be very difficult and time consuming. This military operation is known as ELINT (ELectronic INT elligence) collection. ELINT collection means collecting all radar signals, communication signals, and electronic signals from various sources. The collection can be performed from land, ship, aircraft, or even satellite. A huge amount of data will be collected and processed to obtain the useful information. It might be comparable to gold mining: extracting a small amount of gold from huge amount of sand, gravel, and rock. Not having information about hostile radar can be very costly. During the Vietnam War, whenever North Vietnam obtained a type of missile equipped with new radar from the Soviet Union, the loss of US aircraft spiked [1]. The loss decreased only after the US military learned more about North Vietnam's new radar.

As one can image, the development of radar need be kept strictly secret and, in military operations, radar signals should not be transmitted unless absolutely necessary, especially when a new radar is in use. In some cases, the radar antenna might be replaced by a dummy load and the transmitted signal is absorbed by the load during exercise or testing. Surely this type of practice has its limitations in terms of collecting experimental data for developing new radar. Nevertheless, they make potential enemy's task of collecting information about new weapon radar more difficult.

3.4 Third Layer of Operation: Electronic Countermeasures (ECM)

Once a threat is detected in a battle field by the target aircraft, several operations can be taken to protect the aircraft. These countermeasures may not be all electronic measures. For example, if a heat-seeking missile is launched against an aircraft, the pilot can eject heat sources to distract the missile. These heat sources are referred to as flares. If the missile is guided by a radio signal, one can emit chaffs. Chaffs are made of very thin aluminum foils cut into very narrow strips. The lengths of the strip vary but generally only by a few inches and are comparable to the wavelength of the threat radar signals. The chaffs will be suspended in air for a period of time and reflect radar signals back to the radar. The purpose of the chaff is to create a large number of false targets around the true target in the radar display, thus making the radar lose the accurate target position measurement. These two methods are considered as passive countermeasures because they do not emit signals. They are very important technology and will be discussed in Chapter 7.

In order to track an aircraft and finally bring it down, the radar will go through several steps such as search, acquisition, tracking, and finally launching a missile. If any of these steps are interrupted, the radar may not reach the final step of shooting at the target. It is much better to act against the radar and stop it from obtaining information about the aircraft situation at the early stages. If the radar is sending a missile to the target, the situation is life or death.

Plenty of methods can be used to disturb hostile radar operations. A common approach is to jam the radar by retransmitting the radar signal back to the radar to distract its operation. Since the strength of signal reflected from the target aircraft is rather weak due to the term of $1/r^4$ in radar range equation, Equation (2.1), the jamming signal can be much stronger than the reflected radar signal. Another obvious solution is to emit a noise to mask the true signal on the radar screen. However, as noise covers a broad bandwidth, the energy of noise getting into the radar receiver is limited by the receiver bandwidth. There are numerous approaches for jamming radars and specific technologies are required against specific radars.

All the methods of using signals to interrupt the radar operation are called electronic attack because they emits electromagnetic waves to disrupt the radar operations.

3.5 Third Layer of Operation: Aircraft Tracked by a Radar Can Send a Missile to Destroy the Radar

Aircraft tracked by a hostile radar must protect itself from being destroyed. If the radar is land based, the aircraft might send an air to ground missile to destroy the radar. The missile can use the radar signal as guidance (beacon). This type of missile is called a high-speed anti-radiation missile (HARM). Thus, the radar is vulnerable as well. If a radar is only concerned with offense, that is, to destroy the hostile aircraft but ignore its own protection, it can be easily annihilated by an anti-radiation missile. In order for a radar to protect itself, it must limit its transmitting time. Long operation time may put a radar in a dangerous position. For this reason, the radar should transmit signals only when necessary.

The general rule of any military operation is that both offense and defense must be considered at the same time. In many competitive games such as chess and Go, every move should be made by considering both. For a competitive player, the decision leading to victory or defeat often depends on whether they can be one move ahead of their opponent. The radar and the target aircraft are in a similar situation. One that is one step ahead can destroy the other instead of being destroyed like what is portrayed in the US Air Force Song, "We live in fame or go down in flame. Hey!".

3.6 General Discussion on Second and Third Layer Operation

The second and third layers of electronic warfare operation include intercepting radar signal, identifying signal, attacking the radar electronically or with a missile, etc. It should be noted that all equipment for the second and third layer operation are located on the aircraft. All the operations are almost equally important for the aircraft's survivability. These operations and their relations will be discussed here.

First, let us consider the radar signal interception and classification. Among all the radar signals intercepted, only the threat radars are the top priority. Once the threat signals are identified, an active electronic warfare technique will be applied to disrupt the radar operation. The threat signals being jammed does not interest the intercept receiver. In addition, for reasons to be discussed in following paragraphs, the intercept receiver cannot receive radar signals which are jammed. However, this does not mean that the jammed threat radar signals no longer reach the intercept receiver. The electronic warfare processor just ignores these signals and concentrates on finding new threats. In other words, the intercept receiver should still receive and identify other radar signals to find additional threat radars.

Since the jammer and intercept receiver are co-located in the same aircraft, when a jamming signal is turned on, it will mask the intercept receiver. If the jamming signal is noise, it has a very broad spectrum and can block a big portion of the intercept receiver's working bandwidth. Even if the intercept receiver has a wide working bandwidth, which will be discussed in Chapter 4, a narrow band jamming signal can still potentially block all the receiver operations. For this reason, once a radar signal is jammed, the intercept receiver can no longer see it since the receiver's capability of receiving this signal has been sacrificed by the jamming signal from a colocated jammer.

On the other hand, the jammer cannot stop jamming the radar as long as the radar is in operation. If the jamming signal stops, the radar will operate in normal mode, which can threaten the aircraft. When the jammer is on, the intercept receiver cannot receive the jammed radar signal and it cannot be certain about whether the radar is still operating. If the radar stops transmitting and the jammer still sends out a jamming signal, the jammer wastes its jamming power. Worse yet, a missile sent by the radar can use the jamming signal from the aircraft as a beacon signal. This operation is called home on jamming. As mentioned in previous sections, any equipment that transmits a signal might expose itself to danger. Therefore, when the radar is off air, the jammer must stopping jamming.

In order to check whether the threat radar is still transmitting or not, the jammer must stop periodically to let the intercept receiver determine the radar's situation. As all the information on the radar is already known, this operation is rather simple. The terminology that refers to this operation is called the intercept receiver's look through window. If an aircraft is not jamming a victim radar, its intercept receiver can be fully functional. Because the receiver is a passive device, it does not expose the aircraft. If the jammer is on, the intercept receiver will have limited operation capability.

There are multifarious ways to jam a victim radar. If a radar is jammed by noise, the radar operator will recognize that the radar has been jammed because the noise wipes out the radar display. There is also deceptive jamming, which provides erroneous information to the radar operator. Under this situation, the radar operator may or may not be aware that it is being jammed. The latter operation is sometimes referred to as spoofing.

3.7 Electronic Warfare Aircraft

All of the previous discussions assume that the radar is used against aircraft and the aircraft might attack the radar. The main mission of a fighter aircraft is to attack enemy aircraft or ground targets. Its equipment used against hostile radars is quite limited and is only for self-protection.

Since radar is the main threat to aircraft, there are aircraft built just to operate against hostile radar and they are called electronic warfare aircraft. The equipment on these aircraft can be very sophisticated. Therefore, they are more effective in terms of disturbing radar operations. These aircraft can go into the battlefield and perform all the operations discussed in this chapter up to now except for launching missiles against radars. This practice is more effective because the hostile radar illuminates electronic warfare aircraft directly and they can detect and jam the radar signals through the radar antenna's main lobe. The major risk for an electronic warfare aircraft being at battlefield is that they are defenseless and the enemy can easily attack them.

Another way of operation is to have electronic warfare aircraft stay away from the battlefield and remain outside the adversary's missile range. The electronic warfare aircraft can fly in a certain pattern inside a safe zone. Under this condition, the hostile radar may not detect the electronic warfare aircraft. For example, if the range of the radar is 150 km, an electronic warfare aircraft can detect a hostile radar signal outside the radar's range, say somewhere between 150 and 300 km without being detected (remember the discussion in Section 2.5). In addition, the aircraft may not be in the main lobe of the radar antenna. Because an antenna may have many side lobes, the radar can still receive signals from side lobes. Nevertheless, in order to detect a radar signal in the radar's side lobes, the sensitivity of the intercept receiver must be high. For the same reason, a strong jamming signal is necessary for getting into the radar receiver through radar antenna's side lobes.

3.8 Fourth Layer of Electronic Warfare Operation: Radar Counter-Countermeasure

In military radar designs, all the electronic countermeasures conducted by the targets must be taken into considerations. The radar must detect the jamming signal to avoid providing false information on the target. It can have different operation modes, such as changing its operation frequency, changing the pulse repetition time, etc. These approaches complicate the intercept receiver and electronic warfare processor's task of recognizing a hostile radar. A radar must minimize the illumination time on the target. This is especially important to avoid anti-radiation missiles.

Although a radar with higher power can cover a longer range, high signal power also makes radar vulnerable as an intercept receiver can conveniently detect radars emitting strong signals. One obvious way to avoid being detected is to radiate less power. These types of LPI radar apply just enough power to achieve their operation goal. When they track proximate targets, they reduce their emitting energy. Another way to reduce the probability of being detected can be using a longer low-power pulse such as FM signals, as explained in Section 2.11. By doing so, the total emitted energy remains the same, but peak power is reduced, thus making the radar more difficult to be detected. Other LPI approaches include changing signal frequencies, reducing the antenna side-lobe, using noise-like radar signals, etc. Chapter 9 will introduce some types of LPI radar.

The radar can also send a homing on jamming missile to attack the target. If the target is jamming the radar, the missile can follow the jamming signal. Thus, the jamming signal can bring danger to the jammer. The electronic warfare operations taken by a radar to offset its target's countermeasure are called electronic counter-countermeasure (ECCM) which is also referred to as electronic protection (EP).

3.9 An Electronic Warfare Case Study: Black Cat Squadron

The 1960 U-2 incident, in which the US pilot Francis Gary Powers was shot down near Kosulino while flying a U-2 over the Soviet Union, ignited a major diplomatic crisis in the Eisenhower presidency and is well known to readers in the English world. The lesser known history is that, from 1961 to 1974, the US supplied U-2s to Taiwan's Republic of China Air Force (ROCAF) to conduct surveillance missions over mainland China. ROCAF formed the Black Cat Squadron for this task. As, in the early 1960s, China's People's

Liberation Army Air Force (PLAAF) only had few surface to air missiles (SAM) that could reach the U-2 operational altitude (about 70,000 feet), Black Cat Squadron completed several missions without any loss initially. The Black Cat Squadron even flew over Beijing, taking photographs. It wasn't until September 1962 that the U-2 flew by Major Chen Huai was shot down by PLAAF's SA-2 missile. Afterwards, ROCAF's U-2 was equipped with a radar warning receiver to detect the SA-2 guiding radar signal (fearing it to be lost to China, the US initially did not install a jammer for SA-2 on ROCAF's U-2 until 1964 after losing three U-2s in missions). PLAAF soon noticed that ROCAF's U-2 would change its direction once SA-2's radar was turned on for more than 10 seconds. Based on open sources in Chinese, to handle this situation, PLAAF used the Soviet Union's SON-9 fire director radar for anti-aircraft guns to loosely track the U-2. The U-2's intercept receiver was not designed to detect SON-9 radar (the anti-aircraft gun surely cannot bring down an aircraft at high altitude). The SA-2's radar was switched off until the U-2 was very close to its firing range and the SA-2 was launched quickly (within 8 seconds) after its radar was turned on. By then, it was too late for the U-2 to make any maneuvers. Using this tactic (PLAAF called it *near-fast tactics*), PLAAF brought down more U-2s. The Black Cat Squadron ceased its overflight reconnaissance missions after PLAAF possessed more SA-2s and the whole operation was terminated after the US normalized its relations with China. PLAAF shot down five of ROCAF's U-2s in total.

Some lessons on electronic warfare can be drawn from this sequence of events. First of all, as mentioned before, to avoid being detected, a radar should minimize its illumination time and PLAAF definitely pushed this doctrine to the extreme. One thing worthy of mention is that the U-2 is a defenseless surveillance airplane; so the PLAAF missile unit does not have to be concerned with its own safety and thus can wait until the last minute (or second) to turn on missile guidance radar. This episode of electronic warfare history also demonstrates the importance of ECM. Without a jammer, the U-2 is in a hopeless situation once it is locked by a missile guidance radar. Taiwan did demand the installation of a radar jammer after losing several U-2s and jamming was applied in later missions. Finally, the importance of knowledge about an enemy's radar signal cannot be overestimated. PLAAF used two radars, SON-9 and SA-2's guidance radar to accomplish the task of shooting down U-2s. Being unaware of PLAAF's usage of SON-9 for long-range tracking significantly reduced the U-2's response time.

3.10 Conclusion

The general rule of electronic warfare operations is to limit signal transmission. If enough information is obtained, the emitter should be kept silent because emitting signals can expose the emitter and even attract attacking weapons. Especially for new military radars, the signals must be kept secret. New radar can surprise the hostile aircraft and induce great damage. Intelligence collecting equipment are working all the time. If new radar signals were captured by a rival's intelligence agents, the effect of the radar could be reduced drastically.

In order to protect oneself and act against the enemies in the meantime, the four layers of electronic warfare operations need to continue to improve. The field of electronic warfare is highly technical and keeps evolving constantly. For this reason, it can be a very interesting field for some researchers. In the following chapters, the operation of layers 2, 3, and 4 will be discussed in detail.

References

[1] Mario de Arcangelis, *Electronic Warfare: From the Battle of Tsushima to the Falklands and Lebanon Conflicts*, Blandford Press, 1985.

[2] Bob Bergin, "The Growth of China's Air Defenses: Responding to Covert Overflights, 1949–1974," Studies in Intelligence vol. 57, no. 2 pp. 19–28, Jun. 2013.

[3] Hsichun Mike Hua, "The Black Cat Squadron," Air Power History, vol. 49, no. 1, pp. 4–19, Spring 2002.

[4] Chris Pocock, *The Black Bats: CIA Spy Flights over China from Taiwan 1951–1969*, Schiffer Publishing, 2010.

[5] Lockheed U-2, https://military.wikia.org/wiki/Lockheed_U-2.

[6] 角逐超高空——空軍一支絕密部隊的戰史第三集近快戰法, directed by Ye Liao, CCTV9 documentary, 2018 (https://www.youtube.com/watch?v=-M9TskEpIKw)

4

Intercept Receivers and Electronic Warfare Processors

4.1 Introduction

In a battlefield, an aircraft might be illuminated by many types of radar from both sides. Most radars are benign in nature. However, there are threat radars and their goal is to shoot down hostile aircraft. This chapter describes how to distinguish threat radars from other radars based on their emitting signals. The pulses generated by all the radars are in the order of tens of thousands or even millions per second. From this crowded electromagnetic environment, the threat radars must be found. And not only find them but find them in a timely manner, that is in seconds. If it takes a little longer time such as minutes to identify threat radars, the game might be over. To achieve this goal, intercept receivers and electronic warfare processors are needed.

4.2 Intercept Receiver Requirements

The frequency range of an intercept receiver must cover all the radar frequency. Different types of radar use different frequencies. Usually, the frequency range of military radars is from 2 to 18 GHz. The entire bandwidth is 16 GHz. Let us use an example to show how wide this bandwidth is. The frequency of all FM radio stations is from 88 to 108 MHz and it covers 20 MHz or 0.02 GHz. So, 16 GHz covers 800 FM frequency bands. Within this broad bandwidth, there might be signals from weapon radars. These radars can guide a missile toward an aircraft in seconds. The intercept receiver, electronic warfare processor, and jammer must detect the radar signals, sort them out, identify the threat type, and jam the radar with the correct jamming techniques. All these operations must be finished before the radar can launch a missile toward the aircraft. During the Cold War, Taiwan's U-2 pilots who flew surveillance missions over mainland China were told that they would be safe if they could escape within one minute after the radar warning system on U-2 finds the Chinese threat radar. However, in reality, the window for them to escape was only a few seconds and some brave pilots paid with their lives as the ultimate price.

Besides the need to cover a broad spectrum, the intercept receiver should have high sensitivity. A high sensitivity receiver can receive weak signals; therefore, it can detect radars far away. The other quantity is dynamic range which determines the receiver's capability of receiving simultaneous strong and weak signals. High dynamic range means the receiver can detect both strong and weak signals at the same time. Unfortunately, a receiver cannot have both high sensitivity and high dynamic range. Increasing one value will decrease the other one in basic receiver designs. A compromised decision must be determined by the designer.

In the 1970s, a sales engineer from Watkins Johnson gave an example to describe the requirements of the electronic warfare countermeasures. He said that in a certain city, when the newspaper publishes an interesting news story on a certain day, a certain radio station will broadcast a short question about it. If a listener hears the question, he/she must call the radio station immediately and answer the question. If the answer is correct, the listener will win a grand prize. However, the broadcasting station and the time of the day are not published. In order to win the prize, the listener must buy many radio receivers and tune each one to a certain radio station. Each radio must be operated by a person who listens continuously from the beginning of the day. The number of radios and operators must equal all the broadcasting stations in the city. If the question is broadcast, the listener must call the station immediately. If the answer is correct, he/she will receive the grand prize. If the listener does not provide the correct answer, he/she still will not win the grand prize. Obviously, this is a tremendous effort, and some luck is needed. This is a good analogy to the electronic warfare operation. The grand prize is saving the operators and the aircraft.

The intercept receiver has a very unique requirement, that is, the receiver must be able to receive multiple signals. It is referred to as simultaneous signal receiving capability. In most communication receivers, one receiver is only required to receive one signal. In a battlefield, however, there maybe many different threat radars. The intercept receiver must receive all the radar signals. This is a extremely difficult problem to solve, especially as some signals are weak and some are strong. Some modern intercept receivers are required to receive up to four simultaneous signals.

With this requirement, the radio station example in the previous paragraph should be modified as follows: there can be multiple radio stations broadcasting at the same time. How many radio stations will broadcast similar but slightly different information is unknown but the number is from 1 to 4. The questions asked by the stations are also different. In order to win the grand prize, one must receive all the signals and provide all the correct answers. This is definitely more challenging than the case of one broadcasting station.

4.3 Parameters Measured by Intercept Receivers

What is the information an intercept receiver is required to provide? There are usually five measurable quantities. They are the frequency, pulse width, time of arrival (TOA), pulse amplitude, and angle of arrival. The first four parameters are illustrated in Figure 4.1. The first three quantities are the determined by the radar and the last two parameters are determined by the locations of the radar and the receiver. These parameters are briefly discussed as follows.

The frequency is referred to the frequency of radar signal used to modulate the pulse. This parameter is useful to identify the radar type and provides jammer guidance about which frequency range to jam. The pulse width is the time duration of the radar's pulse signal. It is usually measured in microseconds (μs $= 10^{-6}$ s). This parameter can be used to determine the radar type and provide useful information for jamming. Usually, a weapon radar signal has short pulse width, i.e. 1 μs or below. The TOA is the time the intercept receiver assigns to the time when the pulse is received. This time is not related to any real time but to an internal time generated by the receiver. The information is needed to measure the time between adjacent pulses. The difference between the TOA of two consecutive pulses is the pulse repetition

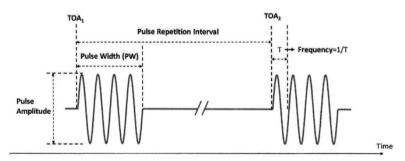

Figure 4.1   Parameters measured by an intercept receiver (angle of arrival not depicted).

interval. This parameter is useful for grouping individual pulses into a pulse train and provides jamming information.

The pulse amplitude is how strong the received pulse is. If the receiver is close to the radar and within the mainlobe of the radar antenna, the signal is strong. This parameter is also valuable for jamming. The angle of arrival is the direction of the radar relative to the receiver. This parameter is extremely useful for sorting the individual pulses into a pulse train and for jamming. The angle of arrival can be considered as the most important information in sorting the signals because modern radar can change the frequency, pulse width, and time of arrival on a pulse-to-pulse basis, but the angle of arrival stays close to a constant. Even an airborne radar cannot change the angle of arrival very quickly. An intercept receiver might receive signals from multiple radars and the angle of arrival can be used to separate signals from different radars. The process of separating signals from different sources is referred to as signal sorting. The angle of arrival also provides the direction of jamming to the jammer. The angle of arrival is usually only measured in the azimuth direction. If the elevation angle is also measured, the distance of the radar can be estimated through the elevation angle and the height of the aircraft if the radar is ground based. The azimuth and elevation angles are illustrated in Figure 4.2. The angle of arrival is the most difficult parameter to obtain because it takes multiple antennas and receivers to measure.

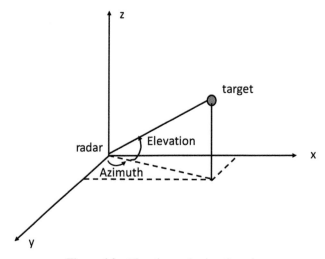

Figure 4.2   Elevation and azimuth angle.

The outputs of a modern intercept receiver are digital words. The words include all the measured information on a pulse-to-pulse basis. Usually, the word is 128 bits long. If an aircraft is flying high in a battlefield, the pulse density is very high. Sometimes, being able to process one million pulses per second is used as a goal for designing an intercept receiver.

4.4 Crystal Radio Receiver

This first field electronic warfare receiver was a crystal video receiver. The crystal video receiver is closely related to the crystal radio receivers which some readers might have built as a science project and they will be introduced here.

When James Tsui was a sixth grader in Qingdao China, the port city famous for its beer, one of his classmates taught him how to build a crystal radio receiver. His classmate acquired this knowledge from his elder brother. At that time, there were only commercial AM radios built with vacuum tubes, but the Tsui family did not own one. James Tsui was addicted to this crystal radio project and invested all of his after-school hours and allowance on it.

There are four major parts in a crystal radio. The two expensive parts are a tunable capacitor used in an old analog radio and a pair of earphones. These two parts could be easily bought from the flea market where plenty of used US military electrical parts were sold at a reasonable price. At that time, the US Navy and Marines were stationed in Qingdao (Qingdao was the headquarter of the Western Pacific Fleet of the US Navy from 1945 to 1949 until Chinese Communists took over China). The other two parts were a coil and a crystal detector. James and his classmate bought old transformers from the flea market to get the insulated copper wire. They then wind insulated copper wires on a section of bamboo to make the coil. The crystal they used was a piece of native copper purchased from a Chinese herbal store (native copper is a traditional Chinese medicine). Native copper is a natural mineral coming in different sizes and shapes. The native copper they bought was roughly a cube with less than 1 cm on each side. The color was dull brown. After being cracked, a shiny metal surface would appear. James learned all these tricks from his friend.

The coil and tunable capacitor of the crystal radio were used as radio tuner tuned to the frequency of a certain AM radio station. The tunable capacitor was used to change receiver frequency. The crystal was used to detect the amplitude of the radio signal and output audio signals to earphones. Nowadays, a diode is often used to replace the crystal. When James Tsui took

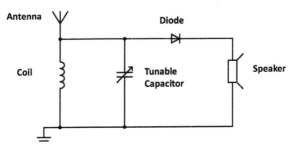

Figure 4.3   A crystal radio schematic.

his first radio receiver to his friend's house for a test, they could hear several radio stations and the sound quality was pretty good because his classmate's elder brother installed a good antenna. But, at James' house, the crystal radio could only receive signals from very few radio stations and the signals were barely audible as James did not have a good antenna at home. Later on, James spent more than 30 years working on electronic warfare receivers in the United States. When he finally went back to visit Qingdao in 1983, he looked for his friend only to sadly find that his childhood friend had moved away and could not be located.

Figure 4.3 shows the schematic of a crystal radio receiver. Note that the diode is used to replace crystal (in James's case, native copper) in this design.

4.5 Electronic Warfare Crystal Video Receivers

Modern electronic warfare can be considered as starting from the Vietnam War. The North Vietnam army obtained air defense equipment from the Soviet Union. Among them, the surface to air missile (SAM) SA-2 is arguably the best known. The radar warning receiver used against the SA-2's radar system was AN/ALR-46, shown in Figure 4.4, which includes a crystal video receiver. The crystal video receiver can be considered the simplest receiver. A simple crystal video receiver diagram is given in Figure 4.5. The incoming signal received by the antenna is passed through a bandpass filter, the filtered microwave signal is then fed to a crystal detector which converts the microwave signal to a video signal, and an amplifier amplifies the video signal.

Microwave frequencies can be divided into different bands based on the size of the waveguide used to transmit/receive signals. The bandwidths of each band can be very wide. The traditional band designation covers the

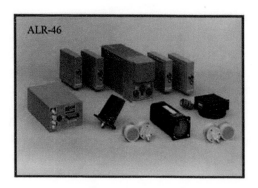

Figure 4.4 AN/ALR-46 system
(https://www.aef.se/Avionik/Artiklar/Motmedel/Nya_hotbilder/RadarWarnStory.pdf).

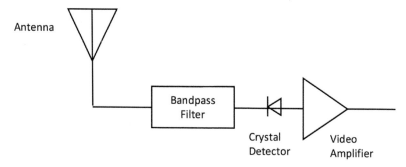

Figure 4.5 A diagram of a crystal video receiver.

spectrum from 2 to 18 GHz in four bands: S (2–4 GHz), C (4–8 GHz), X (8–12 GHz), and Ku (12–18 GHz). One can build a crystal video receiver for each band, and four receivers can, thus, cover the entire radar signal spectrum. If there are four radar signals, each of which falls into one of these four bands, the receiver can detect four simultaneous signals. If, in each quadrant, there are four receivers to cover the whole radar spectrum, the angle of arrival can be measured by comparing the amplitudes of the video outputs from receivers in each quadrant. This arrangement needs a total of 16 receivers. If two radar signals of different frequencies yet in the same band arrive at the receiver at the same time the receiver may produce erroneous data. The receiver can only measure the frequency in very coarse resolution, that is, the bandwidth of the each band.

The major deficiency of the crystal video receiver is its poor measurement accuracy. It can measure frequency and angle of arrival, and has limited capability of detecting simultaneous signals. Its sensitivity and dynamic range are low. Due to issues in its fundamental design, these deficiencies are difficult or impossible to improve. Its major advantages are its size, cost, and wide working bandwidth, which results in a high probability of intercept. Since these receivers are light and compact they easily fit in a jet fighter. During the Vietnam War, these receivers were carried by the US jet fighters including F-4s and F-16s along with jammers and other EW equipment. Since, at that time, electronic warfare was still in its early stages (ALR-46 was the first software-controlled radar warning receiver used in the field), there were only few types of radar in the battlefield and these receivers fulfilled their design goal and protected the aircraft. They are mostly outdated in today's environment.

In the Vietnam War, one of the major threats to US jet fighters/bombers was the SA-2 and the major radar warning receiver used by the US to detect SA-2's radar signals was the ALR-46. These two systems dueled in Vietnam. The SA-2 site launched SA-2 missiles to the US fighters and the US fighters could also launch missiles toward the SA-2 radar site. Both sides suffered losses.

4.6 Superheterodyne (Superhet) Receivers

The superhet receiver was invented by an US engineer Edwin Howard Armstrong in 1918 during World War I. Armstrong also invented FM radio. The superhet receiver was one of the most important inventions in receiver technology. All modern receivers adopt this technology. The frequency of the electrical signal used for communication is relatively high, and, in early times, there was no amplifier for that frequency band available. As a result, the sensitivity of the receiver is low. The idea of the superhet receiver was to change the radio frequency to an intermediate frequency (IF) that is lower than the radio frequency. A simple superhet receiver diagram is depicted in Figure 4.6. The input RF signal is multiplied (mixed) with a signal of different frequency generated by a local oscillator; so its frequency can be shifted to a lower frequency band. Armstrong's contribution was creating a mixer circuit to make such operation possible. The mixer operation can be easily explained by the following equation based on Ptolemy's identities.

$$\sin(2\pi f_1 t) \sin(2\pi f_2 t)$$
$$= 1/2\left(cos2\pi\left((f_1 - f_2)\,t\right) - cos2\pi\left((f_1 + f_2)\,t\right)\right) \qquad (4.1)$$

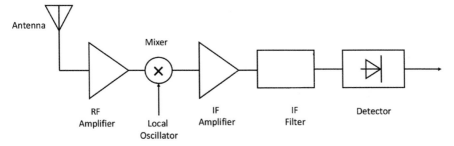

Figure 4.6   A diagram of the superheterodyne (superhet) receiver.

As shown in Equation 4.1, if a device multiplies two sine waves with frequency of f_1 and f_2, it will generate two sine waves: one with a higher frequency of $f_1 + f_2$ and another one with a lower frequency (i.e. IF) of $f_1 - f_2$. The output term with higher frequency can be filtered out and cannot be amplified by an IF amplifier designed to amplify signal at the IF. The beauty of this design is that receivers for different bandwidth can have the same component optimized for the same IF as long as the mixer can down-convert different radio frequencies to the same IF range.

Since amplifiers were available at the IF range. The amplifiers can be used to improve the receiver sensitivity by amplifying weak signals. The frequency coverage of the superhet receiver is in the range of hundreds of MHz. This narrow frequency coverage (the crystal video's bandwidth is the range of GHz) creates three issues. Although the superhet receiver can still cover the whole radar spectrum by continuously changing the frequency of the local signal, it can only cover a relatively narrow spectrum at a time. As a result, the superhet receiver has a low probability of intercepting radar signals. The silver lining is that if a superhet receiver detects a signal, the frequency measurement is rather accurate. Compared with the crystal video receiver, the superhet receiver is more expensive as it needs extra components such as a local oscillator. Also, as the superhet receiver only has one channel, it cannot receive simultaneous signals. It is interesting to note that, despite its technical superiority, the superhet radio receiver was not a commercial success initially due to its cost and difficulty of use.

The superhet receiver has been used in US radar warning systems such as ALR-69 (shown in Figure 4.7) which combines crystal video and superhet receivers. Since the superhet receiver does not cover a wide frequency range, it could not be used or will take a long time to intercept signals in all possible radar signal frequencies. *A priori* information about radar signal was needed

Figure 4.7 ALR-69 system (https://www.radartutorial.eu/19.kartei/12.ecm/karte003.en.html).

for their usage. Although the range of radar frequency is from 2 to 18 GHz, it does not mean threat radars use the whole bandwidth. If the frequencies of the threat radars can be known in advance, the intercept receiver only needs to search the pre-assigned frequency range.

4.7 Channelized Receivers

The idea of the channelized receiver is rather straightforward. Since a superhet receiver with high sensitivity can only cover a narrow bandwidth, building many superhet receivers, each of which covering a small portion of radar signal spectrum (channel), thus covering the whole radar signal band, sounds very reasonable. Obviously, this receiver can be very expensive and have large volume and heavy weight. Does this approach solve most of the intercept receiver problems? The answer is yes and no. This complicated approach creates numerous technical issues and they are difficult to solve. For example, the Defensive Avionics System ALQ-161A installed in the US B-1B used the channelized receiver [1]. This program was not very successful, mainly due to the receiver and electronic warfare processor problems. The original contract of ALQ-161A is 2.5 billion US dollars (in 1982 values), and "it was the largest electronic warfare contract ever signed." Nevertheless, this project experienced many technical difficulties. As a result, the system was delayed, the contract had to be modified, extra budget was spent, and the final product did not meet original promised specifications. For engineers, some solutions look obvious and easy to accomplish. However, when one actually works on it, there might be overwhelming problems and the solutions are no longer obvious. As people often say, "the devil is in the detail."

4.8 Some Common Problems of Intercept Receivers

Intercept receivers have many potential problems in practice originating from their unique requirements. These problems happen to almost all types of intercept receivers. For example, if an intercept receiver is designed to intercept signals within a certain frequency and power range. Then, signals with frequency and power in these specified ranges should be detected correctly. If there is only one input signal, most intercept receivers will produce the correct result. However, even if there are just two input signals, various errors can occur. When two signals with different power are received at different times but their pulses overlap or have close frequencies, detection errors might happen. There are countless possibilities of generating erroneous data and two examples are given here: (1) missing signal: the receiver might miss one or both signals; (2) generating false signals: the receiver might report a signal when there is no such signal at the input of the receiver. People might call the first case false negative and the second case false positive.

The general rule for an electronic warfare system is that it is preferable for the receiver to miss a signal rather than generate false alarms. A missed signal might be detected at a later time, but false information may mislead the electronic warfare processor. The electronic warfare processor takes data generated from the intercept receiver and tries to derive information about radar from it. If the data is inexplicable, the electronic warfare processor will spend valuable time and resources to figure them out. Worse yet, sometimes the electronic warfare processor may produce erroneous radar information based on false data and initiate countermeasure to expose the aircraft. The rest of this chapter will focus on electronic warfare processors.

4.9 Human Electronic Warfare Processor and "SAM-Song"

When James Tsui started to work for the US Air Force Research Laboratory (AFRL) in 1973, his officemate was a second lieutenant named Chris. Chris was an electronic warfare officer. The name of the research laboratory James worked for changed many times during James' employment and the name used here is a generic name which is also its current name. At the beginning, James was in the process of obtaining his security clearance.

Therefore, although Chris and James became friends, Chris never discussed his job. Chris told James some of his experience in Vietnam, and most of them were his off-duty life stories.

One day, James found a black box with an electric cord in the laboratory. He remembered the dimensions of the box as about 12"×12"×4". The box had many toggle switches on the surface. An old-timer told James to plug the electric cord into the wall socket and turn on a switch. When James did, the box produced a very short audio pulse. When James turned on a different switch, a different sound came out. The old-timer said that it was the sound of a crystal receiver when its output is connected to a speaker. He heard one audio output and said that it might be a searching radar's signal. When a certain sound was heard, he said that if an electronic warfare officer heard this sound, he would wet his pants as the sound was the signal from a tracking radar and the threat could be imminent. James did not know whether what the old-timer said was true or not. However, in the Vietnam War, the US pilot could hear a specific tone from their earphones when a Fang Song radar, the missile guidance radar of SA-2, changed from "search" phase to "lock-on" phase and this sound is referred to as the "SAM Song." When the "SAM Song" is heard, missiles were on their way and the pilot needed to make an evasive maneuver quickly. What James heard back then might be the famous "SAM Song."

After James obtained his security clearance and started to work on electronic warfare receivers, one day he thought about the black box and wanted to ask his electronic warfare officer officemate Chris about the box. But, James could no longer find the box because the laboratory was cleaned regularly. Maybe the box was transferred to somewhere else or discarded. The box could be old training equipment. At the beginning of electronic warfare in Vietnam War, the threats were limited; so humans might have some role to play in processing the signals to identify threat radar, and the box James played with was perhaps a vintage piece of equipment from that era.

4.10 Goals of an Electronic Warfare Processor

Joseph Caschera was an engineer at US AFRL. James Tsui met Joe on his first day working at AFRL and they became close friends. Joe was a very quiet and excellent engineer, then working on electronic warfare processor. They went to lunch once a week until James retired in 2004 (Joe retired in 1998). They still kept contact even after James moved to Las Vegas and Joe provided much information about the electronic warfare processor covered in the following sections.

An electronic warfare processor takes the outputs of the intercept receivers as its input and sorts them out. As mentioned in Section 4.3, the radar pulse density can reach one million pulses per second. If each pulse generates 128 bits of data, the amount of information to be processed will be massive. The received pulses might be from many different types of radars received at different instants. An important requirement for the electronic warfare processor is to identify threat radars in seconds. If technology developed for electronic warfare processors cannot meet these requirements, it will just be an academic theory.

4.11 Signal Sorting

From the information of a single pulse, it is difficult to identify the type of radar. It requires many pulses from the same radar to determine the radar type. First, the electronic warfare processors must put all the pulses emitted from one particular radar in a time ordered form. This process is referred to as de-interleaving. To accomplish this goal, the processor can compare the pulse frequencies, pulse width, and pulse repetition interval. These can be considered as the intrinsic properties of the radar. This approach assumes that the frequency and the pulse width of radar signals generated by the radar stays the same from pulse to pulse and the intervals between pulses are fixed. So, radar pulses with the same frequency and pulse width and received at a constant rate can be identified from the same radar. However, not every radar obeys this rule. Some radars can change signal frequency from pulse to pulse and they are referred to as frequency hopping or frequency agile radars. The pulse width can also vary from pulse to pulse. These kinds of practices can be considered as electronic counter-countermeasures (ECCM). Although these radars can be detected by an intercept receiver, they are difficult to sort out. For these radars, the angle of arrival data is the most dependable information. The angle of arrival is not an intrinsic characteristic of the radar; thus, the radar has no control over it.

All information about one radar's pulses such as frequency, pulse width, or angle of arrival need be sorted into one group of data. From the time-of-arrival difference between sequential pulses, the pulse repetition interval can be found. The pulse repetition interval (PRI) and pulse repetition frequency (PRF) are both used in this book. They refer to the same radar feature and PRI=1/PRF. If the radar has only one PRI, the measured value should be only one value. If the radar pulse repetition time is staggered, this radar has several PRI. The measured value should then indicate so. There are also radars that

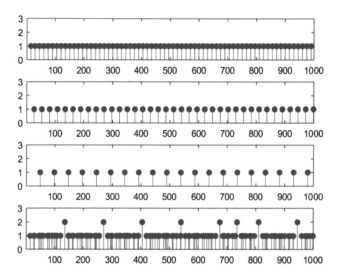

Figure 4.8   Composite signals of three pulse trains generated by three radars. Top three plots: individual pulse trains of three radars, bottom plot: the composite pulse trains. Time unit is arbitrary.

have agile PRI, i.e. the time between sequential pulses is a random number. From these measurements values, frequency, pulse width, and PRI/PRF, the radar can be identified. Figure 4.8 shows the composite signals of three pulse trains, each of which has a constant but different PRI to demonstrate the difficulty of signal sorting. Only few radar signals with constant PRI can produce such confusing results. The results of many types of radars with various PRI will be extremely confusing. If signal de-interleaving is not done correctly, the electronic warfare processor will not be able to extract correct radar features for its identification. As mentioned previously, a wrong radar identification can lead to an ineffective countermeasure which might endanger the pilot's survival.

4.12 Pattern Recognition of Radar Pulses

When James Tsui just joined the AFRL, an old-timer told him that some experienced electronic warfare officers can look at the pulse shape and determine which pulse train they belong to. At that time, the intercept receiver cannot display any radio frequency signals. The only signal that can be displayed is the video signal, that is, what the pulse shape is referred to. He claimed that each magnetron has a special pulse shape and went so far as to claim that from the pulse shape, the age of the magnetron can be estimated as an aged magnetron generates some special leading and trailing pulse edges. If this is true, pattern recognition will be an important technique to identify radars. Pattern recognition of radar pulses was always a subject interesting electronic warfare researchers and it is called pulse fingerprint. Nowadays, people call a similar idea RF fingerprint.

Today, the pulse is collected through digitization of the radio frequency signal. All the detailed information is preserved. In addition, pattern recognition technology has advanced at an amazing speed in past few years. A smart phone can easily recognize a person's fingerprint and it was said that, in China, even vendors on the street can make money transactions through face recognition, not to mention China's ubiquitous security system built upon face recognition. In the authors' opinion, the human face is much more complex than a radio frequency pulse. If a radio frequency pulse can be recognized by the intrinsic characteristic of RF components such as magnetron, antenna, etc., there will be revolutionary changes in electronic warfare. A radar can change frequency and PRI from pulse to pulse, and this practice makes the task of signal sorting more complicated than ever. However, the radar will use the same RF components to generate and emit signals, so this RF signature can be used to group radar pulses. As a result, frequency hopping and changing PRI will not affect the sorting process, provided that the intrinsic characteristic of the RF component does not change during the operation.

The feasibility of this RF signature identification technology will also affect the electronic warfare receiver design. The signal parameters used by future intercept receiver to identify radar may be totally different from what is in use today. This change may even affect the counter-countermeasures. The radar operator might easily differentiate a jamming signal from reflected radar signals as they have different RF signatures.

4.13 Table Looking to Identify the Type of Radar

Once the pulse train emitted by a radar is identified, the data will be compared to an existing file of radars. The file of radars includes the types of a radar and its frequency, pulse width, and PRI. The file should include all radars which might be encountered in the battle. Inaccurate information about new threat radars can bring about a potential disaster. If good intelligence about which types of radar might be found in the battlefield is available, the number of radar in the file can be reduced. In other words, if all possible radars in the battlefield can be known in advance, only these radars are needed in the file. With fewer radar, the searching time can be reduced.

Although commercial search engines like Google have an enormous amount of data and the search time is extremely fast, the electronic warfare processor must have its own special search method to identify the most likely radar from a sequence of radar pulses with missing data and possible errors (and do not forget that it is not impossible that these pulses might be emitted by a radar not on the book). In the early stage of electronic warfare processor development, the search algorithm was one area researchers focused on.

4.14 Trackers in Electronic Warfare Processor

The electronic warfare processor should spend most of its effort on processing new signals. If a pulse train is identified as emitting from a threat radar, the radar should be jammed immediately. If the radar is not a threat radar, the radar can be ignored. Signals from these non-threat radars will still be intercepted by the receiver, but they will be ignored by the electronic warfare processor. Their frequency, pulse width, and angle of arrival data are stored in memory. If new intercepted signals belong to these categories, they will be ignored so that the processor can concentrate on potentially dangerous new signals.

The processor dedicated to perform such functions is called a tracker. Once the data are in the tracker, these data will not be sent to the electronic warfare processor. Therefore, the electronic warfare processor only processes data from unseen radars newly received by the intercept receiver. There are usually multiple trackers.

4.15 Revisit a Signal Being Jammed

If the aircraft is jamming a certain signal, the intercept receiver cannot receive the same signal as the jamming signal is very strong, thus preventing this radar signal from being received. If the intercept receiver is a channelized receiver, theoretically the receiver can intercept signals whose frequency is far from the jamming frequency. In reality, a strong jamming signal with a short pulse width may still block the receiver. If a threat radar stops transmitting but the jammer still works against it, the jammer wastes useful jamming energy. Worst yet, transmitting any RF signals can be dangerous because the enemy's missile can be guided by following the jamming signal; this is called homing on jamming.

In order to avoid this situation, the intercept receiver must revisit the radar signal being jammed. If the radar signal disappears, the jammer should stop jamming. If the signal is still present, the jammer will continue to work. In order for the receiver to revisit the signal, the jammer must stop working momentarily. This operation is called looking through time of the receiver. Since threat radars should be jammed all the time, the looking through time needs to be very brief. Because the frequency of the revisited signal is known, the receiver can be tuned to the desired frequency directly. Therefore, the probability of detecting the revisited signal should be high and a short looking through time will be sufficient. People should not confuse the high probability of detection of a known signal (revisited signal in this case) with the high probability of intercept which means intercepting a new signal. The time of revisiting is determined by the electronic warfare processor.

4.16 Prediction of Pulse Arrival Time

One of the effective jamming methods is called pulse coverage. This method is to send a jamming pulse signal to match one of the radar's returning pulses. Since the jamming pulse is stronger than the returning signal, the radar operator may not recognize the signal received as a jamming signal. In order to mask the returned radar signal, the time of the radar pulse reaching the aircraft must be known. The time can only be predicated by the processor than being measured and the jamming signal is sent at the predicted pulse arrival time. Once the signal is intercepted by the receiver, it is too late to emit the jamming signal.

The TOA of the pulse is measured by the intercept receiver's internal clock. If the radar has constant pulse repetition interval, the next TOA is relatively easy to figure out. If the pulses are staggered, all the pulse repetition intervals must be figured out so that the next pulse arrival time can be predicted. If the radar has a random pulse repetition interval, the pulse arriving time is more difficult to predict.

Once the jamming pulse covers the true skin return, many jamming techniques can be applied. One simple approach is to slowly change the jamming pulse's return time and separate the jamming pulse and the true returning radar pulse. Since the radar follows the jamming pulse, this approach will provide the radar with erroneous distance data. More jamming techniques will be discussed in the next chapter.

4.17 Calculate Radar Crystal Frequency

In all modern radars, there is a crystal oscillator used to generate a time reference. Radar's signal frequency, pulse width, and the pulse repetition frequency are all referenced to the same crystal frequency. Civilian radars usually have stable frequency, constant PRF, and fixed pulse width. Some military radar can change one or more of the three parameters on a pulse-to-pulse basis to make it difficult for the intercept receiver to intercept signals and for an electronic warfare processor to sort radar pulses. Some examples are frequency agile radar, staggered pulse repetition rate radar, and random pulse repetition interval radar. However, none of these parameter changes are totally random. These changes are based on the radar's crystal frequency. All pulse width and pulse repetition intervals should be a multiple of the crystal's period (=1/frequency). Therefore, from the random pulse repetition interval, computer programs can be developed based on the agility of the measured time of arrival to obtain the frequency of the radar crystal. Although the random and staggered pulse repetition frequency makes the task of electronic warfare processor more challenging, it also provides additional information to the electronic warfare processor. The processor can use this additional information to determine the radar's crystal frequency. Since different types of radar use crystals with different frequencies, the crystal frequency can provide supplementary information to sort radar pulses and help to identify the radar type. This is a very interesting situation. A radar can employ some techniques to decrease its probability of being detected, but, on the other hand, these tricks might provide an electronic warfare processor with new information to identify the radar.

In order to obtain the frequency of radar's crystal, the intercept receiver must have fine TOA resolution. The conventional intercept receiver has time resolution of about 50–100 ns (10^{-9} s). The time resolution should improve to 1–10 ns in order to estimate radar's crystal frequency.

4.18 Performance Evaluation of Electronic Warfare Processors

To evaluate an intercept receiver is relatively easy. Usually, there are one-signal and two-signal tests. The procedure is to use one or dual signal generators to generate test signals and change the signal frequency, pulse width, and pulse amplitude. Then read the receiver's output at every test scenario and compare the results with input. The approach should provide a good idea about the receiver performance such as frequency error, amplitude error, pulse width error, probability of intercept (by counting missing pulses), false alarm rate (by counting extra pulses, i.e. the non-existent pulses), etc.

Evaluation of an electronic warfare processor is a much more complicated task. Since the input is radar signals, an intercept receiver must be used to provide the digital words to the electronic warfare processor. In other words, an electronic warfare processor cannot be tested alone but in combination with an intercept receiver. As a result, if the intercept receiver performs poorly, the test results will be poor. Hence, the intercept receiver must have decent performance and, under this condition, the reasonable electronic warfare processor test can be conducted.

The electronic warfare processor is designed for a complicated battle field environment so that a few signal generators will not produce any meaningful results. An electronic warfare processor can be tested with a simulator which generates a certain battle field signal environment. The simulated radar signals are intercepted by an intercept receiver and the receiver outputs are then fed into the electronic warfare processor under test. During the test, the processor reports the types of radar detected and the time used to perform the detections.

A slightly different way to evaluate the intercept receiver and electronic warfare processor is to separate the test into two parts. The first step is to record the outputs of the intercept receiver and store the results. Since the outputs of the intercept receiver are digital words, they can be recorded. These data can be analyzed with a general purpose computer offline to evaluate the performance of the intercept receiver. The second step is to feed these data into the electronic warfare processor and evaluate the results in near real time. This approach can evaluate both equipment separately and predict the overall performance. If the overall system performance is unsatisfactory, this type of test can isolate the problems to be fixed.

4.19 Summary

This chapter discusses both intercept receivers and electronic warfare processors. The major difference between most commercial receivers and an intercept receiver is that the intercept receiver must receive many signals at the same time and might not know the characteristics of signals to be intercepted in advance. A few intercept receivers are introduced. The outputs of the intercept receivers are also discussed. These output data can be fed into a general purpose computer to evaluate the performance of the intercept receiver.

If the intercept receiver outputs are obtained from a complex battlefield electromagnetic environment, the parameters of intercepted signals such as frequency, pulse width, and angle of arrival can be sent to an electronic warfare processor, and the processor needs to sort them out into pulse trains and compare them with a library of parameters of different radars to identify the threat radars. The library of radars is gathered through electronic intelligent collection operations. It takes tremendous amount of time and many different operations such as collecting from aircraft, intelligence collection ships, etc., to fill the library.

Due to advances in pattern recognition, radar signal pulses may be sorted out by checking RF signatures of radar pulses. If this method is successful, the entire field of electronic warfare can be revolutionized. Not only new electronic warfare processors are needed, but the requirements for an intercept receiver will also change. Thus, new electronic warfare receivers can emerge. Furthermore, pattern recognition may also be used by radars as counter-countermeasures to improve the radar's performance against jamming. The duel between electronic warfare systems and radars probably will never end and any new technology available will likely be used by both sides to improve their chance of survival.

References

[1] Alfred Price, *War in the Fourth Dimension*, p. 74 and p. 175, Greenhill Books, 2001.
[2] Alfred Price, *History of US Electronic Warfare, vol. 3*, The Association of Old Crows, 1984.
[3] Edgar O'Ballance,"The impact of European Armies of the United States Vietnam Experience," in the Proc. of the 1982 International Military History Symposium: The Impact of Unsuccessful Military Campaigns on Military Institutions, 1860–1980.
[4] Mario de Arcangelis, *Electronic Warfare: From the Battle of Tsushima to the Falklands and Lebanon Conflicts,* Blandford Press, 1985
[5] James Tsui, *Digital Techniques for Wideband Receivers*, 2nd edition, SciTech Publishing, 2004.

5

Jamming and Counter-Countermeasures

5.1 Introduction

Once a signal emitted by a threat radar is detected, in order to protect the aircraft action must be taken immediately against the threat radar. The required response time is in the order of seconds. The required action against the radar is usually called jamming. The jamming is a general term and different jamming techniques are applied against different types of radar.

Usually, one intercept receiver is used to receive radar signals from all different types of radar for the electronic warfare processor to identify. When the electronic warfare processor recognizes the radar type, this information is sent to a technique generator and the technique generator will produce the corresponding jamming signal. The jamming signal must be amplified and sent to the threat radar. Some radar can be jammed easily, but some are more difficult to interrupt. In some cases, one jamming source may not accomplish the jamming goal alone. It might take several jammers working together simultaneously and these jammers can be at different locations. This fact shows the difficulty of conducting effective jamming.

Since the practice of jamming is to send a signal/noise back to the radar, it is also called active electronic warfare operation in contrast to the passive electronic warfare used for intercepting signals. Another common term used for this practice is electronic attack (EA). The aircraft dedicated to active electronic warfare are also called electronic attack aircraft. People not familiar with electronic warfare might associate electronic attack with some fictional weapons in movies, but it is not what electronic attack stands for. Perhaps, the phrase EA was invented because of the military's appetite for the word "attack."

The main goal of jamming is to protect the aircraft. Thus, weapon radars are the primary target. The weapon radar must first lock onto a target aircraft to obtain the necessary information before launching a missile against it.

71

When the target aircraft is locked onto by a weapon radar, the attack can be imminent. The goal of jamming is to break the lock. When the radar breaks lock, it can no longer provide information on the aircraft and thus cannot attack it. The weapon radar must try to lock onto the target aircraft again. This operation is time-consuming and the radar might not relock on the target.

The action against jamming referred to as electronic counter-countermeasures (ECCM) could be included in a separate chapter. However, just like many jamming schemes are designed against specific radars, many ECCM were developed to against certain jamming methods. In this book, selected jamming techniques are described along with their advantages and disadvantages. In the authors' opinion, instead of dedicating a separate chapter to ECCM, it might be easier to introduce corresponding ECCM techniques when discussing the shortcomings of certain jamming techniques. Otherwise, if ECCM is discussed in a separate chapter, these jamming techniques must be addressed again in order to refresh the readers' memory. For this reason, the ECCM will also be discussed in this chapter.

One major advantage that an EW system possesses over a radar is that the jamming signals only travel half the distance reflected radar signals propagate through to reach the radar. To successfully jam or deceive a radar, it is desired to have a jamming signal stronger than the reflected radar signal when both of them reach the radar. For this reason, despite the shorter transmission, the jamming signal needs to be amplified appropriately. We live in a world of solid state devices which are used in almost every electronic device. Nevertheless, while solid state microwave generators are widely used in the laboratory, they usually cannot deliver the power required by radar. Therefore, a magnetron is still used to generate radar signals in many cases. In a similar situation, although solid state microwave amplifiers have been widely adopted for different applications, they usually still cannot generate enough power for the jammer. To provide high microwave power, traveling wave tube amplifiers (TWTs) invented in the 1940s are still in use. For this reason, a brief introduction to TWTs will be given in the next section.

5.2 Traveling Wave Tube (TWT) [1, 2]

A traveling wave tube (TWT) is a special vacuum tube used to amplify microwave signals. The first TWT related device was invented by Andrei "Andy" Haeff in 1931, but Rudolf Kompfner is often credited with inventing TWTs in 1942. TWTs are a very mature technology and their basic circuit has changed very little since the 1940s. While solid state devices have replaced vacuum tubes in almost every application, TWT is still commonly used for satellite communication, radar, and electronic warfare applications due to its broad working bandwidth, high efficiency, huge amplification gain (typical value: 60dB, i.e. 1,000,000), and small footprint.

Figure 5.1 shows an illustration of a TWT. The TWT consists of three parts: an electron beam, a slow-wave structure, and a magnetic shield. The electron is emitted by a heated cathode and accelerated by the voltage applied between cathode and anode toward the far end of the vacuum tube. A magnetic shield is outside of the tube which contains the electron beam. The purpose of the magnetic shield is to focus electrons into an electron beam and keep the beam at the center of the tube. Outside the electron beam are microwave slow-wave structures. There are two types of slow-wave structure: helix and cavity. The TWT shown in Figure 5.1 is a helix TWT. The microwave signal entering the slow-wave structure is slowed down so that its speed along the electron beam propagation direction is comparable with the electron beam speed. As a result, the electron beam and the slowed down microwave can interact, and, through this interaction, the energy of the electron beam is transferred to the microwave signal and the microwave is amplified.

It appears that building a TWT is a pretty difficult technical challenge; yet, its applications are limited in few specialized fields. If there is a breakthrough in solid state microwave devices, the TWT might follow the footsteps of other

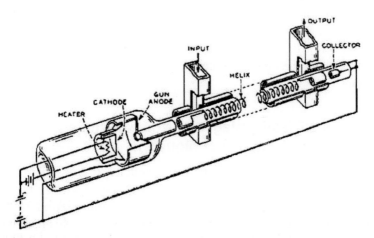

Figure 5.1 Picture of a helix traveling wave tube (https://en.wikipedia.org/wiki/Traveling-wave_tube#/media/File:Traveling_wave_tube_diagram.png).

types of vacuum tubes and become a part of history. Nevertheless, the military still needs these tubes currently. Therefore, once in a while, James Tsui would hear the following conversation among his coworkers in the Air Force. It is difficult to train new technicians or engineers in the field of TWT. It will take a person very long time to learn all the tricks in this specialized and mature field. So, James' coworkers joked that a special program should be set up to recruit people to learn how to make TWTs and offer them a life-time job even if, one day, the tube is no longer needed. This kind of talk shows the importance of TWTs to the military.

5.3 Noise Jamming

Noise jamming is a common experience in everyday life. Most of time, the noise is not generated intentionally. For example, when someone talks over the phone, if there is significant noise on the line, people usually hang up and redial or even ask the person on the other side to hang up and call back. Hopefully, the noise will go away through this practice. Background noise is one of the most difficult problems in communications. Under various conditions, noise is used to prevent the listeners from getting information. For example, if a foreign government does not like its people to listen to Voice of America (VOA), it can generate noise to interrupt broadcast in certain frequency range such that the VOA broadcasting is difficult to recognize.

To use noise to jam a radar, the jammer needs to generate a noisy signal whose spectrum covers the radar's operating frequency and send it toward the radar. If the power of noise received by the radar receiver is strong enough, the noise can cause the radar screen to be fuzzy and the reflected radar signal will be hidden below the noise. All being well, this operation will make the radar lose its lock on the target. Figure 5.2 shows a radar screen masked by noise jamming.

5.4 Advantages/Disadvantages of Noise Jamming and Burning Through Techniques

The advantage of noise jamming lies in its simplicity. It does not require detailed information about the radar signal. For noise jamming to be effective, the jammer needs to make sure that the jamming noise received by the radar carries more energy than the reflected radar signal. For this reason, if possible, the jammer should focus its energy in a narrow spectrum around

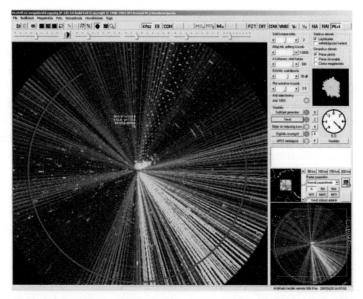

Figure 5.2 Radar screen masked by jamming noise (https://www.radartutorial.eu/16.eccm/ja01.en.html).

the radar's operation frequency as noise outside radar receiver's working spectrum will be filtered out, thus having no jamming effects. If the frequency of the radar signal is known, the jammer can generate a noisy signal in the same frequency range and send it toward the radar. Other than radar's operation frequency, no other information about the radar signal is needed.

By definition, a noise is a wideband signal covering a broad bandwidth.[1] To limit the noise spectrum, a bandpass filter centered at the specific frequency can be used to filter out the portion of the noise outside the desired bandwidth generated by a noise generator, thus creating a band-limited noise. Or, a low-frequency noise can modulate a radio wave and the resulting noise will be centered at the frequency of the radio wave and the noise bandwidth is equivalent to the original noise bandwidth. The band-limited noise is often

---

[1]The spectrum of a noise is an advanced engineering topic. In most undergraduate engineering curricula, the spectrum/bandwidth of a noise is not covered until senior year. Roughly speaking, the spectrum of noise is determined based on how well one can estimate future/past noise value based on the current noise value. If the current noise value reveals nothing about any future/past noise value, then this noise is a white noise whose spectrum is infinite. The description about noise in this section is, of course, not theoretically rigorous, but description about colored noise implementation is technically sound.

referred to as colored noise. The term comes from the optical spectrum. When the light of all frequencies (or wavelengths, which are the reciprocal of the frequencies) are present, the color of light is white. If only a certain frequency band of light is present, the light has a certain color. The term white noise, meaning the noise spectrum is from zero to infinity, is so well recognized that it was used as the title of a horror film in 2005. To jam a radar, the jammer generates colored noise, amplifies it, and then sends it toward the radar to disturb the radar's operation. Not only the idea of noise jamming is simple, the operation is also simple. There are several types of noise jamming and some of them are introduced below.

Barrage Jamming. Barrage jamming applies noise across a broad bandwidth to jam the radar. Due to the usage of broadband noise, barrage jamming is capable of simultaneously jamming multiple radars, each of which might have different operating frequencies, and the jammer does not necessarily know the exact operating frequency of the radar to be jammed. The major disadvantage of barrage jamming is that as barrage jamming covers a broad spectrum, only a portion of noise is received by the radar. As a result, to overpower reflected radar signals, the jammer needs to send out a very strong jamming noise.

Spot Jamming. Unlike barrage jamming, spot jamming concentrates all of the noise energy in a narrow spectrum. This method requires knowledge about the radar's operating frequency. If this information is available, then the jamming signal can easily overpower the reflected radar signal as the energy of jamming signal is proportional to $(1/R^2)$ and the energy of the reflected radar signal is proportional to $(1/R^4)$, where R is the distance between the target and radar. As the spot jammer concentrates all of its energy in a narrow bandwidth, to jam multiple radars, multiple jammers will be needed. If the radar's frequency is unknown the chance of a spot jammer successfully jamming a radar will be small.

Sweep Spot Jamming. Barrage jamming covers a broad bandwidth, so its chance of jamming a radar is high, but most of the jamming power will be wasted. As a result, its jamming might not be effective unless a very high power jammer is utilized. On the other hand, spot jamming can be very power efficient if the spectrum of jamming noise matches the radar's operating bandwidth. However, for a radar with unknown frequency, the chance of successful jamming is small. Sweep spot jamming tries to solve this problem by sending a narrow band jamming noise whose frequency is swept through a broad spectrum. As a result, the chance is that many radars will be jammed at some point although not always. The hope is that the frequent disturbance caused by the sweep spot jammer will be enough to make radar lose its target.

Simple operations usually have some limitations. One disadvantage of noise jamming is that the radar operator knows the radar is being jammed. Simply from the display, the radar operator can recognize that the radar has been jammed by a noise source. This information is very important for the operator to make the correct decision. The use of noise to jam radar operation is not as effective as its use for jamming a communication system. The communication signals are not repetitive, but many pulse radars keep sending out the same signal so that the reflected radar signal is repetitive. The display on a radar screen is the integration of many returned signals. It is difficult, if not impossible, to derive useful information based on one returned signal.

The returned radar signal is coherent (i.e. the phase of signal is fixed), but the noise is incoherent. If we add coherent signals, the total signal strength increases. Of course, if we add noise, the amplitude of combined noise also increases. However, the rate of increase for the signals and noise is different. To explain this phenomenon, consider adding aligned sinusoidal waves together; in this case, peaks are added to peaks; and valleys are added to valleys. As a result, we will get a sinusoidal wave with a larger amplitude. Now, consider adding many noises together. Since noise is incoherent, there are no fixed phase relations between noises; so we will not always add peaks to peaks and valleys to valleys. Consequently, the total signal strength increases faster than the total noise strength increases in this process. For this reason, if the radar operator keeps adding returned signals, the returned signal may appear again on the radar screen. This operation is sometimes referred to as burning through. Figure 5.3 demonstrates burning through. The left side of Figure 5.3 shows that a sine wave is seriously corrupted by noise and it is difficult to see the sine wave from the corrupted signal. However, after adding five corrupted signals together, the result shows a very noticeable sine wave, as illustrated on the right side of Figure 5.3.

It should be mentioned that many EW professionals refer to burn through as a range at which the energy of the reflected radar signal from the target is larger than the energy of the jamming signal received by the radar. As a radar signal usually has energy higher than the jamming signal's, although the reflected radar signal energy is proportional to $1/R^4$ and the received jamming signal energy is proportional to $1/R^2$, if R is small enough the reflected radar signal energy will be larger than the jamming signal energy. Within the burn through range, the noise jamming will no longer be effective.

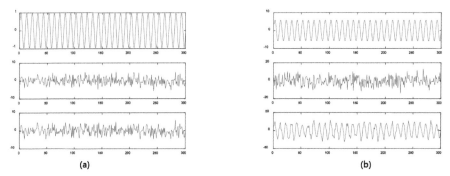

Figure 5.3 Demonstration of burning through. (a) Top: one signal, middle: one noise, bottom: sum of top plot and middle plot. (b) Top: sum of five signals, middle: sum of five noises, bottom: sum of top plot and middle plot.

5.5 Deceptive Jamming and Cover Pulse Operation

The goal of deceptive jamming (also known as spoofing) is to deceive the radar with erroneous information. Hopefully, the radar operator does not notice that the radar is being jammed and makes a wrong decision based on incorrect information. The logic, of course, is that the operator's wrong decision will provide the tracked aircraft with a better chance of accomplishing its mission. One obvious advantage of spoofing is that the power required for the jammer is not as demanding as the noise jammer. However, this technique is more sophisticated. A major device to perform the spoofing is a technique generator. It is used to generate the desired jamming signals.

A radar can measure three major parameters of its target aircraft: the *distance* of the target, the *angle* of the target with respect to the radar, and the *speed* of the target. Let us do a quick review of how radar determines these three parameters. A pulse radar periodically sends out a short pulse of microwave signal and then listens to the reflected radar signal (skin return) to find its target. Based on when the radar receives the reflected signal, the radar can determine the distance of the object. In the easiest implementation, each transmitted pulse is a short microwave with a fixed frequency. The frequency of the reflected signal can be different from the frequency of the transmitted signal due to the Doppler effect discussed in Chapter 2. Using this frequency difference, the radar can then determine the target's velocity. When searching for an object, the radar's antenna will scan though the sky and the direction from which the radar receives a strong reflected signal tells the radar the angle of the object. This operation is illustrated in Figure 5.4.

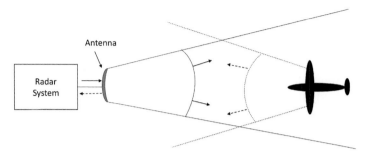

Figure 5.4 The working principle of pulse radar.

A successful spoofing requires jamming a radar without alarming the radar operator. To accomplish this goal, the jammer needs to make the radar operator mistake the jamming signal for an authentic skin return pulse. First, the jamming signal must take over the radar's return signal without causing any obvious change on the radar screen. To do so, the jammer needs to generate a jamming signal to match the radar signal. The timing of sending out a jamming signal is also important. If the jammer receives the radar signal and immediately sends a jamming signal, this spoofing will not be successful because there is always a time delay between intercepting a radar signal and generating a jamming signal. In order to send a jamming signal at a correct instant, the jammer must predict the next radar pulse's time of arrival and, when this time comes, sends the jamming signal whose time duration covers the duration of the radar pulse so that the radar receiver receives the skin return and the jamming signal simultaneously. The pulse predication is the task of the technique generator. Because the jamming pulse covering the radar returning signal has similar characteristics, the radar operator may not notice the existence of the jamming pulse and the jamming pulses then take over the operation of the radar. Once this task is accomplished, the radar's estimation of the object's distance, speed, and angle can be misled by the jammer.

It is worthy of notice that there is a variant of noise jamming which uses a noise pulse to cover the radar's skin return pulse. With this approach, a noise jammer might jam multiple radars, and the requirement for jammer's power can be relaxed. This jamming method is referred to as cover noise jamming. Cover noise jamming is not a deceptive jamming as it does not provide radar with wrong information. In the following sections, we will discuss some deceptive jamming techniques used to alter one or more of the radar's three measurements and the radar's counter-countermeasures.

5.6 Range and Velocity Deception: Range Gate Pull-off and Velocity Gate Pull-off

When a pulse radar detects an object via the reflected radar signal, it will put a range gate around the time the skin return pulse is received and null all of signals received outside the window. The reason behind this practice is to reject unwanted noise (and a jamming signal sent out at a wrong time). Of course, this range gate will move along with the reflected radar signals as the time the reflected signal is received is determined by object's distance. When a target aircraft's EW receiver intercepts this radar signal, it needs to figure out the signal's frequency, pulse width, and pulse repetition interval. Once it determines the radar pulse repetition interval, it can send a jamming signal with pulse width and frequency similar to the radar's pulse (this task can be accomplished by sampling the intercepted radar signal and then duplicating it) around the time when the next radar pulse reaches the target aircraft. Initially, the jamming signal and reflected radar signal roughly overlap, but, gradually, the jamming signal is separated from the skin return. Usually, the jamming signal is stronger than the true radar returning signal, so the range gate will follow the jamming signal instead of the real return signal and eventually the range gate only receives jamming signal instead of the true reflected signal. If the jamming signal gradually changes its transmitting time, such as moving ahead of the radar pulse, the radar will indicate that the target is getting closer. If the transmitting time is moving after the radar pulse, the radar will show the target is getting farther away. This deceiving jamming technique is called range gate pull-off. This operation is illustrated in Figure 5.5.

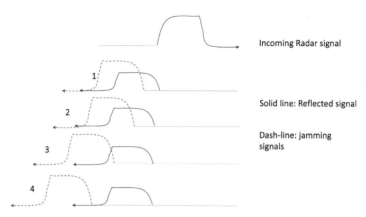

Figure 5.5   Range gate pull-off.

In a similar way, the frequency of the jamming signal can also change gradually. A small change in frequency can provide the wrong velocity information because the radar measures the frequency shift from the returning pulse to determine the Doppler frequency and the target's speed. This method is called velocity gate pull-off. A radar might use a bandpass filter (velocity gate) around the frequency of its skin return to track the target's speed and filter out noise. The jammer initially sends a strong jamming signal with the same frequency as the skin return radar signal's frequency and then gradually changes the jamming signal's frequency to move the radar's velocity gate so that it filters out the actual skin return radar signal. After successfully moving the range gate or velocity gate of the radar far enough so that these gates filter out the actual skin return radar signals, the jammer can stop transmitting the jamming signal and the radar will then lose the track of its object and needs to search for it again. It is worthy of note that velocity gate pull-off can be applied against both pulse and CW radars.

5.7 Electronic Counter-Countermeasure: Frequency Agile and Pulse Agile Radars

As mentioned before, an intercept receiver measures five parameters of radar pulses: the frequency, time of arrival, pulse width, pulse amplitude, and angle of arrival. The first three parameters are the intrinsic radar signal parameters which the radar can control. In order to increase the radar sensitivity, a radar detects its object based on several received skin return pulses. Civilian radars usually keep these three parameters constant so that it is easier to integrate received skin return pulses.

On the other hand, due to anti-jamming considerations, for military radars, its frequency, pulse width, and time of arrival can be altered on a pulse-to-pulse basis. Of course, these modifications will complicate the radar design. For example, if a radar changes its signal frequency on a pulse-by-pulse basis, not only the transmitter design is difficult but the receiver design is also complicated. A radar transmitting signals with different frequencies is referred to as a frequency agile radar. Radar can also have many pulse repetition intervals (staggered PRI) which will create different intervals between times of arrival on the intercept receiver side. Radars with a staggered PRI possess PRI agility. Similar operation can be performed on the pulse width.

As mentioned in the previous section, the first step of the deceptive jamming is to cover the radar pulses with a jamming signal so that the jammer can take over the radar's operation. To cover pulses, the technique generator must be able to predict the next radar pulse's frequency, pulse width, and time of arrival. If some of the radar intrinsic parameters are changed on a pulse-to-pulse basis, it becomes difficult to predict the next pulse's parameters. With all these parameters continuously changing, it will be challenging for the intercept receiver and electronic warfare processor to measure and de-interleave radar pulses, let alone predict the next radar pulse. Therefore, many military radars are parameter agile in nature to counter deceptive jamming.

5.8 Angle Deception: Sidelobe Jammer and Sidelobe Cancelation Scheme

The jamming can also cheat the radar operator on angle measurement. Like all antennas, the radar antenna has a mainlobe and many sidelobes, as shown in Figure 5.6. When searching for its target, the radar emits radar pulses, the majority of whose energy is through the mainlobe. When detecting a reflected radar pulse, the radar receiver assumes that the energy it receives is through antenna's mainlobe and the direction of the mainlobe is the direction of the target. To deceive a radar with the wrong angle information, the jammer avoids sending the jamming signal when the main beam of the radar illuminates the target. When the main beam is not pointing at the target, the jamming signal is sent to the radar through the sidelobes of its antenna. Since the antenna gain of the sidelobe is much weaker than the mainlobe,

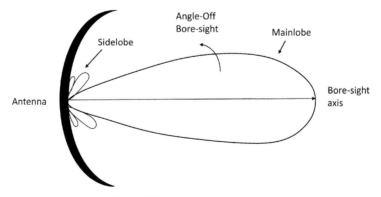

Figure 5.6   Radar antenna pattern.

the jamming signal must be rather strong. The radar usually considers that all the signals are from the mainlobe. If the radar determines the jamming signal is the true skin return, it will decide that the target is in the mainlobe direction, which is the wrong direction. Thus, the radar may provide erroneous angle information.

The issue of receiving signals through the antenna's sidelobe is a general antenna problem not unique to radar. The intercept receiver has the same issue. One of the important radar parameters the intercept receiver needs to measure is the signal's angle of arrival. This information is not an intrinsic radar parameter and the radar operator cannot control it. Thus, this becomes the most critical parameter measured by the intercept receiver. For an electronic warfare receiver, the angle information can be used for two important operations. The first one is radar pulse sorting. Since the radar cannot control its angle of arrival, the angle of arrival is the most reliable parameter for sorting radar pulses. In this operation, the true value of the angle of arrival is not critical as long as the measured values are consistent and can be used to separate radar pulses from different radars. The second usage of angle of arrival information is to direct the jamming direction. In this operation, the correct angle of arrival value is required. If the angle measurement is wrong, the jammer will point in the wrong direction to perform jamming. An intercept receiver measures the angle of arrival with its antenna which also has a mainlobe and sidelobes. Radar antennas are usually well designed to increase their directional gain so that the signals transmitted/received through sidelobes are much weaker than the signals transmitted/received through the mainlobe. On the other hand, the antenna of an intercept receiver is simple, light weight, and small so that it can be installed on a jet fighter. As a result, this antenna has relatively high sidelobes. Although no one tries to build a jammer to jam an intercept receiver, the receiver itself can generate erroneous angle of arrival information due to the shortcomings of its simple antenna design.

One common way for a radar to eliminate the possibility of producing wrong angle information from antenna's sidelobe is to use the sidelobe cancelation scheme illustrated in Figure 5.7. The same concept can be applied to both the radar and intercept receiver systems. The idea is to introduce another antenna and receiver. The second antenna pattern is omni-directional which means the antenna has equal gains from all directions. The requirement for the second antenna is that its gain from all directions is stronger than the sidelobe gain of the main antenna but weaker than the main antenna's mainlobe gain. When the output of the main receiver is stronger than the second receiver's output, it is determined that the signal is from main antenna's

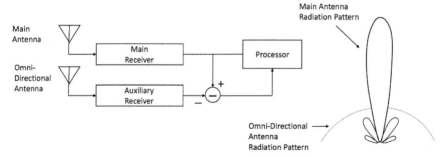

Figure 5.7   Sidelobe cancelation scheme.

mainlobe and the measured angle is correct. Otherwise, the signal is from the sidelobe of the main antenna. With this design, the radar operator knows whether the radar is being jammed deceptively by wrong angle information.

5.9 Angle Deception: Inverse Gain Jamming

From the name, one can guess that this jamming is based on the signal strength measured. The intercept receiver measures the intercepted radar pulse amplitude and uses this one parameter to control the technique generator. Of course, the jammer must have the correct frequency and time of arrival. In other words, the jammer already covers the radar pulses.

The idea behind inverse gain jamming is that when the mainlobe of the radar antenna points at the target, the intercept receiver receives a strong signal. When the antenna's mainlobe points to somewhere else and its sidelobe points to the target, the intercepted radar signal has much weaker energy. To exploit this phenomenon, when the measured pulse amplitude is high, a weak jamming signal is sent to the radar. When the measured pulse amplitude is weak, a stronger jamming signal is sent to the radar. That is where the name inverse gain comes from. As a result, the inverse gain jamming can disturb the radar's angle measurement because the radar then receives strong return signals over a broad range of angles. If the jamming signal takes over the radar, the strength of the returned signal is controlled by the jammer instead of the true radar returning signal.

This jamming scheme is very effective against the conical scan radar covered in Section 2.13. Since the mainlobe of an antenna is relatively broad, the direction measured is not accurate. The conical scan radar overcomes this problem by making its antenna do a cone-shaped scan to put the target at the center of the scan. The technical approach is accomplished through comparing the outputs of the antenna. When the outputs of the scanning antenna have about the same amplitudes through the scan, the target is at the center of the cone. If not, the radar can adjust the center of cone scan to locate the target. The inverse gain jamming disturbs this practice. As a result, the antenna cannot position the target at the center of the conical scan antenna.

The design logic of the inverse gain method is that, during the conical scan, the radar antenna emits a signal while scanning and uses the skin return radar pulse's fluctuating energy during the scan to precisely locate its target; therefore, the inverse gain is conceived to "equalize" this energy variation. So the corresponding radar's counter-countermeasure is to have one fixed radar antenna point in one direction and have a separate receiving antenna performing the conical scan. This way, the intercept receiver cannot observe energy variations during the scan and inverse gain jamming will not work. This ECCM is referred to as COSRO (Conical Scan on Receive Only) or LORO (Lobe on Receive Only).

5.10 Angle Deception: Cross-eye Jamming [3, 4]

As mentioned in Section 2.14, a monopulse radar uses 4 antennas to transmit/receive signals and relies on the sum and difference of these antennas' received signals to locate its target. As a result, it can locate its target with a single pulse and this is how this radar got its name. The advantage of the monopulse radar is the difficulty of jamming it. For example, inverse gain jamming described in the previous section actually helps the monopulse radar improve its sensitivity as it can use the strong jamming signal received by all the antennas to locate the target.

One jamming method designed to jam monopulse radar jamming is cross-eye jamming. This basic concept of cross-eye jamming is to use two separated jammers at different locations sending jamming signals which are out of phase by 180°. In general, these two jammers can be installed at the tips of the wing of the aircraft since it is the farthest separation an aircraft can have. In the ideal case, these two jamming signals are received by different antennas of a monopulse radar. As they are out of phase, when they are added together, they cancel out each other and the radar sees nothing, and when one received signal subtracts another one, the radar sees the maximum output. Although the ideal case is very unlikely, cross-eye jamming can result in the output of the sum being smaller than the output of difference. Resultantly, the radar will tune its angle away from the target and eventually lose its lock on the target.

From this simple description, one can see that this operation is rather difficult because the phase difference of two jamming signals needs to be carefully controlled over a broad bandwidth. One straightforward implementation is having two jammers on the tip of each wing and each jammer has one receiver and one transmitter, as illustrated in Figure 5.8. Each jammer's transmitter rebroadcasts the received radar signals from the other jammer's receiver and one of the two jamming signals is delayed to create the 180° difference. However, the amount of delay depends on signal frequency (for a 2-GHz signal, the delay needs to be 0.25 ns and for a 1 GHz signal, the delay is 0.5 ns) and the jammer has no control over the radar's frequency. For the cross-eye jamming to be effective, this delay (phase difference) needs to be accurate. This is an extremely challenging task. The original idea of cross-eye jamming was filed for patents in 1958 (US4117484A and US4006478A), but the actual deployment did not happen until the 1970s (these two patent applications were granted in 1978 and 1977, respectively). The complexity of this clever approach might be the main reason behind its delayed deployment.

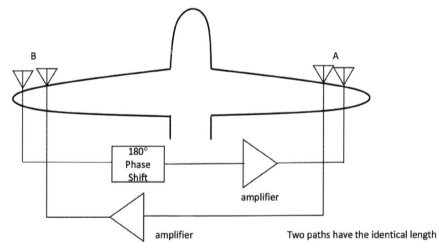

Figure 5.8   The cross-eye jammer implementation concept (adapted from Journal of Electronic Defense, April 2010, p. 50).

5.11 Digital Radio Frequency Memory (DRFM)

As can be seen from previous descriptions of different deceptive jamming techniques, to fool the radar, the jammer needs to generate a jamming signal very similar to the radar's signals. With the advances in radar, this task becomes more and more difficult as there are so many possible radar waveforms. Even if the radar only sends out a pulse signal at different frequencies, it is still difficult for the jammer to synthesize a fake radar skin return in a short period of time, not to mention the radar might change its pulse width, RPI, and frequency on a pulse-to-pulse basis. One way to solve this issue is to record useful radar pulses that the intercept receiver receives, store them in memory, and regenerate these pulses if necessary. Digital radio frequency memory (DRFM) was developed for this purpose.

The earliest reference to DRFM can be found in 1975 and it has become a standard device for jamming. The basic idea of DRFM is very simple; it down-converts received radio frequency signal to a lower frequency and then uses a high speed analog-to-digital converter to sample the resulting low frequency signals. The principle of frequency down-converting a signal is the same as the working principle of superheterodyne (Superhet) receivers covered in Section 4.6. If the sampling rate is high and the receiver bandwidth is wide enough, then a high fidelity replica of a received radar pulse can be obtained and stored in the memory. The DRFM can regenerate radar pulses stored in memory or modify them for deceptive jamming purpose and then up-convert these pulses (using the same principle of down-converting, i.e. multiplying the signal by a radio wave generated by a local oscillator) back to radio frequency for transmission. A simple diagram of DRFM can be found in Figure 5.9. The DRFM is even capable of storing a train of radar pulses with different pulse widths and PRIs so that it can jam a radar with pulse agility. With rapid advances in digital devices, DRFM can only be more versatile in the future and will be an indispensable device for the electronic warfare industry for a very long time.

5.12 Electronic Warfare Aircraft

The jamming techniques covered in this chapter can be performed by either the target aircraft (self-protection jamming) or by an escorting electronic attack aircraft (support jamming). The support jamming can be further classified as stand-off jamming or stand-in jamming. The stand-off jamming is conducted by an electronic attack aircraft behind the target aircraft and the

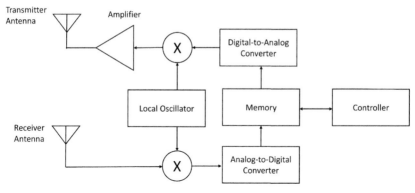

Figure 5.9 DRFM diagram.

stand-in jamming is performed by a jamming aircraft (probably remotely piloted) between the target airplane and the radar. The relations between self-protection and escort jamming and between stand-off jamming and stand-in jamming are illustrated in Figure 5.10. For self-protection jamming, the target aircraft can carry an electronic countermeasure (ECM) pod. Figure 5.11 is a photo of ECM-pod AN-ALQ 101. The main mission of an electronic attack aircraft such as the Northrop Grumman EA-6B Prowler and the Boeing EA-18G Growler is to disturb radar operation and communications and these aircraft usually have all the necessary electronic warfare gear to perform effective jamming.

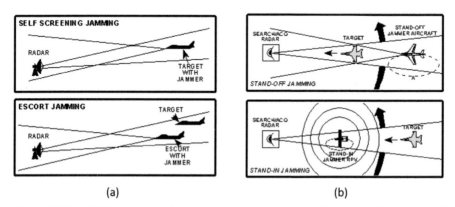

(a)   (b)

Figure 5.10 (a) Self-protection jamming and escort jamming. (b) Stand-off jamming and stand-in jamming [5, 6].

Figure 5.11   AN-ALQ 101 ECM pod.

During a battle, sometimes aircraft equipped with EW systems fly first to attract the enemy radars to find or shoot at them. For example, in the Vietnam War, after losing aircraft to North Vietnam's SA-2 missiles but not being able to locate SA-2 missile sites and destroy them, the concept of Wild Weasel was conceived. The idea was to have Wild Weasel units that consist of aircraft equipped with a radar warning receiver to lead other aircraft in a mission. The function of the Wild Weasel unit was to attract enemy radars to lock on Wild Weasel units and the Wild Weasel plane could destroy the enemy radar site by using anti-radiation missiles, which can follow a radar's signals. This is, of course, a very risky job. From June 1966 to August 1966, out of 10 Weasel-modified F-105Fs (the second batch of Wild Weasels), 5 of them were lost.

In a modern war, without effective radars, the military is literately blind. On the other hand, without an effective ECM, the aircraft are sitting ducks. During war, the main job of the electronic attack aircraft is to protect all friendly aircraft. They usually fly in a certain pattern away from the immediate battlefield and can jam the enemy radar at a far distance. Since they are equipped with better electronic warfare equipment, they can concentrate on jamming. One such example is that, on December 18, 1972, in a mission to attack Hanoi, besides aircraft conducting actual attacking tasks, 31 EA aircraft were deployed for stand-off jamming.

Support jamming has some obvious advantages. When an EA aircraft deceptively jams the radar with false angle information, the radar will have huge difficulty in distinguishing between the real skin return (or the skin return the radar wants to see) and jamming signals because the jamming signal actually provides correct angle information of the EA aircraft. However, the radar cannot see the skin return radar pulse reflected from the target aircraft it is looking for. Another obvious advantage is that the EA aircraft is safe from homing on a jamming missile because it is outside of the battlefield. Therefore, the jammer has little or no threat from the adversary.

5.13 Dangers of Jamming

All active electronic warfare operations are inherently dangerous because emitting radio signals can potentially be self-exposing and jamming, especially self-protection jamming, is no exception. Some missile guidance modes are designed to follow the jamming signals. They are referred to as homing on jamming. The missile can be guided by either a guidance radar or its own radar. When a missile is guided by a radar, the radar is sending a guiding signal aimed at the target aircraft and the missile follows the reflected radar signals to approach the target.

If the missile is equipped with its own radar, it will transmit a signal toward the target aircraft and follow the skin return radar signal. After the aircraft jams the radar or the missile, the missile can change its operation mode. Instead of following its own guidance signal, the missile can change its mode to homing on jamming and use the jamming signal as the guidance. The anti-radiation missile uses the same principle to attack radar.

5.14 Battle Examples: EW in the Yom Kippur War and the 1982 Lebanon War [11–14]

On October 6, 1973, the Jewish Day of Atonement, also known as Yom Kippur, the holiest day of the year in Judaism, Egypt and Syria, launched a coordinated surprise attack on Israel. In the south, Israeli airbase in the occupied Sinai Peninsula was attacked by the Egyptian Air Force and the Syrian Air Force attacked Israeli defense installation on the Golan Heights in the north. Before the attack, Arabs jammed Israeli radio communications causing chaos and confusion in the Israeli chain of command. Nevertheless, after the initial stage of confusion, the Israelis struck back, sending Phantoms and Skyhawk to attack Egypt. Being experienced from previous conflicts with

Egypt, the Israeli military was confident about its air superiority. However, unknown to the Israeli Air Force, Egypt had received the new missile system SA-6 from the Soviet Union before the war. Unlike the pulse radar based SA-2 and SA-3 systems the Western World were familiar with, the SA-6 uses the low-power *Straight Flush* CW radar to illuminate target airplanes. The intercept receiver installed on Israeli jet fighters was designed to intercept pulse radar, and thus failed to pick up the low power CW radar signal, let alone jam it. In addition, the *Gun Dish* radar used to control the Egyptian anti-aircraft guns also changed its radar signal frequency to a much higher frequency. As a result, Israel lost 100 jets in single week and asked for help from the US. The US delivered ALQ-119 jamming pods which use the velocity gate pull-off method covered in Section 5.6 against CW radars. Nevertheless, as so little was known about SA-6, neither US nor Israel knew if ALQ-119 would deliver the promised performance. As a result, Israel used ALQ-119 only against SA-2 and SA-3 but not SA-6.

Despite its air force's heavy losses, Israeli ground force was able to hold its ground on two fronts and even advanced. A cease-fire was then declared on October 24, 1973. Learning from this expensive lesson, in 1982, when another fight between Israel and the Palestine Liberation Army/Syria erupted in Lebanon (known as the 1982 Lebanon War), Syria's SA-6 sites in Bekaa Valley were Israel's main targets. After collecting intelligence about the Syrian radar's operation parameters using electronic warfare aircraft, Israel used high speed drones as bait for Syrian missiles. On these drones was installed electronic warfare equipment which can generate a response to the Syrian radar similar to the one from a jet fighter. Subsequently, Syrian radars started to search for these drones and, following the Syrian radars' signals, Israeli jet fighters carrying jamming pods initiated an attack on Syrian missile batteries. Israel claimed that, within 15 minutes, 17 out of 19 Syrian SA-6 sites were put out of action.

Several lessons can be learned from these two historic events. First of all, different radars require different jamming techniques and the importance of intelligence about radars cannot be overemphasized. Actively sending signals can be dangerous. Fooled by Israeli's drones, Syrian's SA-6 radars were on for a long time to search for these drones which eventually led to their destruction. Most jamming techniques covered in this chapter were developed against pulse radar but low-power CW radar (it can spread energy over a long period of time) is hard to detect and can be effective. A more detailed introduction to a variant of CW radar (FMCW radar) will be given in Chapter 9.

5.15 Conclusions

This chapter discusses a few jamming techniques, their advantages/ disadvantages, and corresponding counter-countermeasures. Some real battle examples are used to demonstrate important electronic warfare principles. Jamming is so important that the electronic warfare system of an aircraft spends more time on jamming than on receiving. That is why there is a look through time for the intercept receiver. The look through time is designed for the jammer to stop jamming for a very short time so that the intercept receiver can check whether the signal being jammed is still in the air. As the jamming signal can reveal the aircraft and be used by the enemy as a missile guidance signal, once the jammed radar signal disappears, the jamming signal should also be stopped.

Deceptive jamming is better than noise jamming. If the operation is carried out smoothly, the radar operator may not recognize that the radar has been jammed until it loses track of its target. The primary requirement of deceptive jamming is to know the radar signal to be jammed. Deceptive jamming starts by covering the radar's returning signals with a jamming signal. If the radar signal is designed with some parameter agility, the jammer will have a difficult time covering the radar's pulse. Some radars are difficult to jam and special jamming techniques are required. In recent years, DRFM has become a major EW device that has enabled EW systems to jam complex radars.

If the target aircraft fails to jam a radar and a missile is launched toward the aircraft, it needs to outmaneuver the missile or mislead the missile to survive. This will be the topic of the next chapter.

References

[1] Samuel Liao, *Microwave Devices and Circuits*, 3rd edition, Prentice-Hall, 1990.
[2] Damien Minenna, Frédéric André, Yves Elskens, Jean-François Auboin, and Fabrice Doveil, "The Traveling-Wave Tube in the History of Telecommunication," the European Physical Journal H, Springer, vol. 44, no. 1, pp. 1–36, 2019.
[3] Warren P. du Plessis, "Practical Implications of Recent Cross-Eye Jamming Research," in the Proc. of Defense Oper. Appl. Symp. (SIGE), São José dos Campos, Brazil, 2012, pp. 167–174..

[4] L. Falk, "Cross-Eye Jamming of Monopulse Radar," in the Proc. of 2007 International Waveform Diversity and Design Conference, Pisa, 2007, pp. 209–213.

[5] *Electronic Warfare Fundamentals*, U.S. Department Of Defense, 2000.

[6] *Electronic Warfare and Radar Systems Engineering Handbook*, 4th edition, Naval Air Warfare Center Weapons Division, 2013.

[7] David L. Adamy, *EW 104: EW Against a New Generation of Threats*, Artech House, 2015.

[8] Andrea De Martino, *Introduction to Modern EW Systems*, 2nd Edition, Artech House, 2018.

[9] S. J. Roome, "Digital Radio Frequency Memory," Electronics & Communication Engineering Journal, vol. 2, no. 4, pp. 147–153, Aug. 1990.

[10] Sheldon C. Spector, "A Coherent Microwave Memory Using Digital Storage: The Loopless Memory Loop," the Journal of Electronic Defense, Jan./Feb., 1975.

[11] Alfred Price, *War in the Fourth Dimension*, Greenhill Books, 2001.

[12] Alfred Price, *History of US Electronic Warfare, vol. 3*, The Association of Old Crows, 2000

[13] Mario de Arcangelis, *Electronic Warfare: From the Battle of Tsushima to the Falklands and Lebanon Conflicts*, Blandford Press, 1985.

[14] David Eshel, "EW in Yom Kippur War," the Journal of Electronic Defense, pp. 49–55, Oct. 2007.

6

Missile Detection Schemes and Defense

6.1 Introduction

The major focus of Chapter 4 is intercepting methods for determining whether an aircraft is illuminated by a threat radar. If a threat radar is detected, the radar must be jammed at once, and Chapter 5 focuses on jamming and radars' counter-countermeasures. In a modern battlefield, there are hundreds of thousands of pulses per second. So, there is always a chance that the threat radar is missed by the intercept receiver and electronic warfare processor. In addition, some missiles can be launched without using a radar. As a result, a missile is launched toward the aircraft and the pilot needs to take immediate action to save their life. In this chapter, we will focus on missile detection schemes and defense.

It is worthy of notice that missiles are not the only threat to the aircraft. Anti-aircraft artillery (AAA) can also be a major threat. As a matter of fact, during the Vietnam War, the US lost more jet fighters to AAA than to surface to air missiles (SAM). Even in the Operation Desert Storm, the US lost 9 aircraft to AAA compared with 10 lost to radar-guided SAM and 13 to infrared (IR) guided SAM. Many AAA are controlled by a radar; so the intercepting and jamming methods covered in the previous two chapters can improve a pilot's survivability against AAA. Nevertheless, after a shell or rocket is fired, since it is unguided, its motion is governed by its momentum and gravity. As a shell's direction cannot be changed (but its fuze might be jammed) and the best (and probably the only) way to avoid a shell is to fly away from it; so there is not much to address about how to defend an aircraft from AAA, and this chapter will focus on how to detect and defeat a missile.

By definition, the major difference between a missile and a rocket is that a missile is guided and a rocket is not. However, not every missile needs radar for guidance. For example, some heat secking missiles like an IR SAM can be fired by an infantry soldier through visual guidance and the missile follows the aircraft using heat radiated from the aircraft. Under this condition, traditional passive electronic warfare techniques for detecting radar signals cannot detect the missile and warn the pilot. Therefore, the radar defense system cannot solve all of the issues about detecting missiles. If a missile is launched and is on its way to the target aircraft, but the pilot and the electronic warfare officer of the aircraft are not aware of this urgent danger, one can imagine how devastating the situation will be.

Because of the this possibility, a missile approach warning system not based on intercepting radar signals is needed. Two approaches to detect a flying missile will be presented in this chapter. The first approach is passive and the second approach is active.

The passive approach is to detect the plume of the missile. When a missile is in flight, its engine needs to push it forward. The engine is hot and has an exhaust plume behind it. Theoretically, if one can detect the plume, it can detect the missile. Since the spectrum of the plume is in the infrared range, this missile approach warning system is called an IR-based missile approach warning system. Certainly, an IR guided missile applies the same principle to follow its target aircraft. The active missile approach warning system is a simple radar used to detect an approaching object. Of all the threats, a missile approaching the tail of the aircraft is probably the most dangerous one. Therefore, this radar is looking backward from the tail of the aircraft to find any approaching object. For this reason, this radar is named the tail warning radar.

The passive approach is relatively simple and cost effective. An IR missile approach warning system can be installed in many different kinds of aircraft including a helicopter. On the other hand, the active approach is more complicated and relatively more expensive. This approach might be suitable only for larger aircraft such as bombers.

6.2 Basic Concept of Infrared Receivers

Infrared is the electromagnetic wave whose wavelength is between 0.7 and 1000 μm (10^{-6}m). Its spectrum is between visible light and microwave. The infrared spectrum can be further divided into five ranges: near-infrared (0.75–1.4 μm), short-wavelength infrared (1.4–3 μm), mid-wavelength infrared (3–8 μm), long-wavelength infrared (8–15 μm), and far-infrared (15–1000 μm). The spectrum of infrared is illustrated in Figure 6.1. Based on the Stefan–Boltzmann law, any object with temperature (T) above absolute zero emits radiation and the total amount of radiation is proportional to T^4.

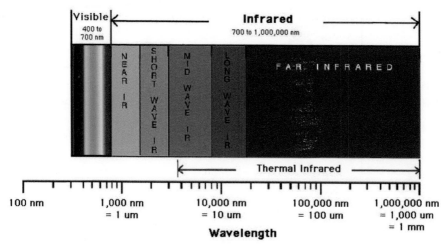

Figure 6.1 Infrared spectrum (https://sites.google.com/a/globalsystemsscience.org/digital-earth-watch/key-messages/near-infrared-and-the-electromagnetic-spectrum).

As 99% of thermal radiation is in the infrared spectrum, infrared radiation from an object with a heat source such as a missile or an aircraft is a great signature used to detect these objects. It is the reason why an IR guided missile is often referred to as the heat seeker. The obvious infrared radiation source of a missile is its exhaust plume and hot engine, but a fast-moving missile can heatup its surrounding air; so this aerodynamic heating can also generate infrared radiation.

The IR missile approach warning receiver is often placed on the tail of an aircraft to detect an approaching missile, but for a helicopter, IR missile approach warning receivers might be installed in its four quadrants. When a missile is launched, its exhaust plume is very strong. If the missile can be detected at its launching stage, the pilot can have maximum response time. For a missile being "cold" launched, which means that the missile is ejected into the air first and its engine is then ignited, the targeted aircraft's response time might be shortened. After a missile is detected, the IR-based missile approach warning system can follow its trajectory.

Theoretically, an IR-based missile approach warning system is a great idea and its working principle cannot be simpler. However, one major problem with this approach is other infrared radiation in the environment especially radiation from the sun. The sun is much hotter than most of the heat sources on Earth. Not only is the sun brighter, it also shines over many places. Sun rays can be reflected from different places toward the IR receiver

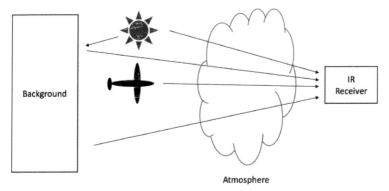

Figure 6.2 Infrared radiation reaching the receiver.

and the receiver might report approaching missiles due to this natural thermal radiation. In addition, infrared radiation can be absorbed in the atmosphere, which is weather dependent. This situation is illustrated in Figure 6.2. So, it is an engineering challenge to make a reliable IR-based missile approach warning system and the issues of IR-based missile approach warning systems will be discussed in the next section.

6.3 False Alarm Problems and Other Issues of the IR-Based Missile Approach Warning System

One of the major issues with an IR-based missile approach warning system is its false alarm. Unlike the infrared sensor installed in a missile which goes after a large object like an aircraft, the IR-based missile approach warning system is designed to detect a small object in a short time. Therefore, it needs to have a high sensitivity. On the other hand, high sensitivity brings about a high false alarm rate. An infrared receiver tested in the laboratory may function very satisfactorily, but, in reality, during flight tests, it can generate many false alarms. A long time ago, someone once told James Tsui about the performance of IR-based missile approach warning systems. He said the first thing a pilot did was to turn off the warning system because it generates many warning signals and the pilot did not know how to handle them. So, for pilots, the best approach to deal with this issue was to turn off the warning system. Of course, this conversation happened a long time ago and, with advances in technology, this statement is probably no longer true. Nevertheless, compared with other missile approach warning systems, the IR-based warning system's high false alarm rate remains one of its major disadvantages.

Most people have the experience of driving toward the sun. The sunlight can be unbearable for some drivers. Everything in the direction of the sun is difficult to see clearly. In the old days of air dogfights, one of the common tactics of pilots was to attack from the direction of the sun. After IR air-to-air missile is widely deployed, how to use sun or its reflections to confuse IR missiles is an important maneuver to learn. Compared to the sun, the radiation from a missile's exhaust plume is weak, especially when the missile is in flight. After James Tsui discussed with his coworker who worked on infrared receivers about infrared receiver's false alarm problems, he started to pay attention to the reflection of the sun. When he traveled in an airplane, he always looked down to see the reflection of the sun from the ground and he could see all kinds of reflections, i.e. from a small pond, buildings, etc. Most of the time, he did not even know where reflections were from, but many of them were very bright.

Up to this chapter, all of receivers discussed in this book are radio receivers. For people working on radio frequency intercept receivers, their major efforts are to improve a receiver's performance and false alarms are not a serious problem. When testing radio receiver, false alarms are seldom an issue because the radar signals are specifically generated and they do not exist in nature.

Although the sun generates a tremendous amount of heat and, thus, infrared radiation, the infrared radiation from a missile exhaust plume has special spectrum. Different heat sources might have different infrared radiation, as shown in Figure 6.3. This information can be used to distinguish between infrared radiation from a missile and other objects. Moreover, even different types of missiles can have different infrared spectra. For example, some data shows that the infrared radiation from a vehicle-launched surface-to-air missile is in the range of near or short-wavelength infrared and the infrared radiation from all surface-to-air missiles and air-to-air missiles can be found in the range of mid-wavelength infrared. The infrared signatures of missiles can be very useful to reduce the IR receiver's false alarm rate. Some approaches use filters to filter out certain portions of the infrared spectrum. If the outputs from these optical filters match a missile's infrared radiation spectrum, a missile is then detected. Some approaches check two different infrared sub-spectrums to find a missile. In reality, it is difficult to measure and identify the entire spectrum of the exhaust plume of every possible air threat. In order to develop IR missile approach warning systems, a huge amount of exhaust plume data are needed, and James Tsui's coworker friend went on many trips collecting them. All of these efforts were spent to reduce

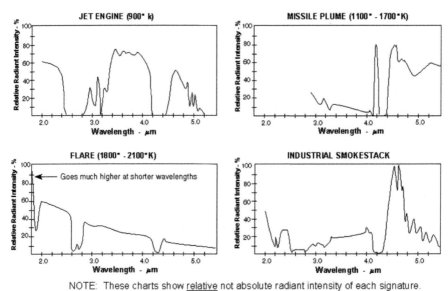

NOTE: These charts show <u>relative</u> not absolute radiant intensity of each signature. Consequently the "amplitude" of one cannot be compared with the "amplitude" of another.

Figure 6.3   IR radiation spectrum of different targets [7].

the IR missile approach warning system's false alarms so that it can be used to detect the true incoming missile.

Besides a high false alarm rate, another issue of an IR missile approach warning system is the atmosphere's absorption of infrared radiation. As shown in Figure 6.2, before any infrared radiation reaches a receiver, it passes through atmosphere and the infrared radiation is absorbed during this propagation. Making things even more complicated, this absorption depends on air composition. Different types of molecules absorb infrared radiation of different wavelengths and the air composition is altitude dependent. So, when infrared spectrum is measured and used for detecting an approaching missile, two things need to be considered: the radiation spectrum and the atmospheric absorption spectrum. In addition, water is a very strong infrared absorber, thus the infrared receiver does not function well in bad or foggy weather.

Because of these issues, some researchers have looked at light of different wavelengths for missile detection, and ultra-violet (UV) radiation is also used to detect approaching missiles. For such a system, its photo detector is designed to only detect ultra-violet radiation on the shorter wavelength side to avoid UV radiation from the sun (referred to as solar-blind UV). Most of the sun's UV radiation with wavelength shorter than 280 nm is absorbed by the Earth's atmosphere. As the UV and IR-based missile approaches warning systems are designed based on the same principle and the major difference between them is wavelength; the UV receiver will not be discussed in this book.

6.4 Tail Warning Radar

The tail warning radar is designed to detect missile approaching aircraft from behind. The requirement of the radar is relatively simple, but it has to be light and small. Another design challenge is that the approaching missile is usually small and has a very small radar cross section (RCS) which is the area reflecting radar signals, thus making it difficult to be detected. A missile might be long, but it usually has a small diameter. When a missile is flying toward the radar, the cross section facing the radar can be tiny, and, during the missile's flight path, its head may be always facing the tail warning radar. If the missile and the radar are not aligned in the same line, the RCS of the missile might be larger.

The range of the tail warning radar depends on from where a missile might be launched. The missile can be fired from a ground base or from an aircraft. Consequently, the tail warning radar should cover a relatively large angle, not only looking directly backward.

One major advantage of the radar-based missile approach warning system over the IR-based one is that it does not generate false alarms as frequently as the IR-based missile approach warning system. A tail warning radar can also estimate the approaching missile's speed and range. On the other hand, the tail warning radar sends out signal which can expose the aircraft, thus increasing the vulnerability of the aircraft. A tail warning radar also might not detect a missile with a small RCS until it is too late. In order to reduce the false alarm rate of the IR-based missile approach warning system, combining the tail warning radar and IR receiver has been considered, but it might not be a feasible option for every type of aircraft.

6.5 Proximity Fuze and Jamming

The first time James Tsui heard about the idea of a proximity fuze was when he was a college student. His professor talked about German anti-aircraft artillery shells and emphasized two technologies used in the shell. The first one was that the wall of the shell is part of a storage battery but without the electrolytic fluid. The electrolytic fluid is in an ampule in the shell but not in the battery. When the shell is fired, the ampule is broken by the shock, the shell spins, the centrifugal force pushes the fluid into the battery, and the battery is then activated. The battery is used to power a simple radar. When the battery is active, the radar is tuned on. If the received signal amplitude is high, it means that an object is close, and the shell explodes. James Tsui totally forgot about why his professor talked about this subject. Although the concept of proximity fuze was described, the word proximity was never mentioned. At that time, James was wondering how expensive the shell was. Later on, he realized that military operations can be extremely costly.

Actually, the proximity fuze was an active research topic among several countries before and during the World War II (it probably still is). The proximity fuze based on a radio signal was first designed by Britain engineers and shared with the US in the early stages of World War II (in the same trip, the British showed the Americans how to make a magnetron for generating radar signals). The US worked to perfect the fuze design, and the clever battery activation scheme described in the previous paragraph is a US invention (the implementation is different though; the wall of shell is not part of the battery, which is a stand-alone component). A transmitter sends out a continuous wave signal, the same antenna is used for both transmission and receiving, and a receiver detects a Doppler signal generated by combining the reflected signal and the local oscillator's voltage. The working principle is the same as the superheterodyne receivers covered in Section 4.6. The Doppler signal's frequency is the Doppler shift caused by velocity difference between shell and target. The Doppler signal's frequency is much lower than transmitted signal's frequency and its amplitude is proportional to the reflected signal's amplitude. When the target is near, the reflected signal's magnitude increases. If the Doppler signal's amplitude is above the threshold, the fuze detonates the shell. This fuze was named the VT fuze in which VT stands for variable time. A diagram of a proximity fuze is provided in Figure 6.4. Although the same principle can be used for bomb and other artillery, this fuze is only used in the antiaircraft artillery of naval ships at first because the chance of an anti-aircraft shell directly hitting an aircraft is scant, and, in this practice, the chance of a proximity fuze falling into the enemy's hands is also limited. The VT fuze was not authorized for ground warfare until in 1944 and General George S. Patton was very impressed after seeing its use in the Battle of the Bulge. As for the fuze's cost, the cost per fuze was $732 in 1942 (2020 value: $11,617) and $18 in 1945 (2020 value: $258.68).

The working principle of the proximity fuze of modern missiles is similar to the one just described, although the mechanism of powering the radar

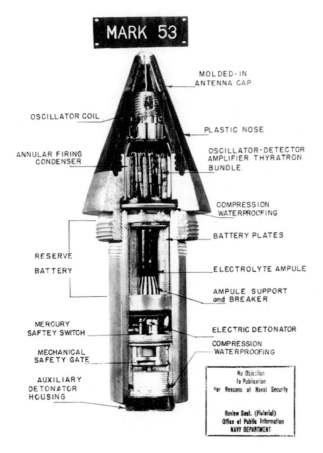

MARK 53

MOLDED-IN ANTENNA CAP

OSCILLATOR COIL

PLASTIC NOSE

ANNULAR FIRING CONDENSER

OSCILLATOR-DETECTOR AMPLIFIER THYRATRON BUNDLE

COMPRESSION WATERPROOFING

BATTERY PLATES

RESERVE BATTERY

ELECTROLYTE AMPULE

AMPULE SUPPORT and BREAKER

MERCURY SAFTEY SWITCH

ELECTRIC DETONATOR

COMPRESSION WATERPROOFING

MECHANICAL SAFETY GATE

AUXILIARY DETONATOR HOUSING

No Objection
To Publication
For Reasons of Naval Security

Review Seal. (Pictorial)
Office of Public Information
NAVY DEPARTMENT

Figure 6.4 Proximity fuze (https://commons.wikimedia.org/wiki/File:MK53_fuze.jpg).

might be different. Since the proximity fuze uses a radar system, it is subject to jamming. The purpose of jamming is to cause a premature detonation when the missile is still far away from the target aircraft. One simple way to jam a proximity fuze is to amplify the received radar signal and retransmit it so that if the proximity fuze is activated based on the amplitude of Doppler signal, the jamming signal can cause a premature detonation. To counter this repeater jamming method, some fuzes detonate the shell only when the Doppler signal's amplitude is above the threshold and the Doppler signal's frequency changes rapidly. The second condition is added based on the physics phenomenon that when the missile flies toward the target, the Doppler shift is positive and when the missile flies away from the target, the Doppler shift is negative. Different radar signals have been used for proximity fuze radar as counter-countermeasures. As we already know, to effectively jam any radar, the importance of knowledge about radar signals can never be overestimated.

6.6 Missile Guidance Systems

There are two common missile guidance systems: radio frequency and infrared. The radio frequency guidance systems can be further divided into two types: one with radio guidance signals originating from the missile itself (referred to active homing) and one with radio guidance signals originating from a radar (referred to as semi-active homing). To jam either active homing or semi-active homing missiles, the guidance signal must be known. If one does not know the missile's guidance signal, one can be surprised in the battlefield and suffer a great loss, as shown by the example of the Yom Kippur War described in Section 5.14 in which the Israeli Air Force lost one hundred aircraft in a week due to a new missile, SA-6, which was unknown to them at that time. Once the guidance signal is known, the loss can be significantly reduced, as demonstrated by the example of the 1982 Lebanon War also covered in Section 5.14.

Another popular missile guidance approach is through infrared sensors. The missile is designed to follow the heat emitted from the aircraft which results in infrared radiation. The aircraft's infrared radiation can be from (1) its exhaust plume, (2) its engine heat, and (3) the airplane's frame (reflection of sun or heat generated through aerodynamic heating). The earlier IR missiles were designed to home in on the hot engine part (near-infrared band) and later heat seekers follow the engine exhaust plume (mid-infrared band). The IR guiding system is a type of passive homing system in which the missile does not emit signals. Home on jamming which uses aircraft's jamming signal for guidance is also a type of passive homing system. Some of the IR missiles are designed to launch without any radar guidance. An operator can aim the IR SAM at an aircraft visually and launch the missile. One example is the US's stinger missile, one of many man-portable air defense systems (MANPADS). This kind of missile guidance is also called fire-and-forget because all the missile needs is someone to fire it. Once the missile is launched, there is no guidance signal needed; so the person firing it can leave/hide and the missile will automatically seek and follow the target.

The IR guidance system is a photonic system with IR detectors to detect infrared radiation from the target. Its operation principles are similar

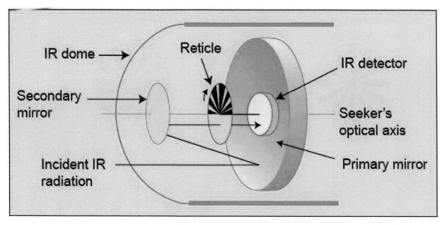

Figure 6.5 Basic infrared guidance system (https://www.electronicsforu.com/technology-trends/precision-guided-munitions-infra-red-guided-weapons-part-3-4/2/).

to IR-based missile approach warning systems. The guidance system needs to differentiate between the target and other infrared radiation sources and cover a wide range of angle. Its tracking angle also needs to be accurate. A diagram of a guidance system is shown in Figure 6.5. In this system, the infrared light passes through the IR dome, reflected by the primary mirror and then by the secondary mirror, passes a rotating reticle whose function will be explained in great detail later, and then reaches the IR director which generates voltages based on radiation intensity. The reticle can be considered as a circular lens with a certain pattern of transparent and opaque areas. Its purpose is to help the guidance system to differentiate between target and other background IR radiation source such as cloud or sand and provide accurate angle information about the target. To achieve these goals, different reticle patterns are used and some of them are shown in Figure 6.6. Reticle A is called the full spokes reticle, and it has interleaving transparent/opaque spokes. A full spokes reticle can differentiate between background radiation and a target's radiation. Assuming the target size is small and background IR sources such as cloud are large, when the reticle rotates, the background IR source creates a relative constant voltage at the receiver and the target IR source creates a periodic voltage (like a periodic pulse signal) at the receiver. Also, when the target is closer to the center of the reticle, the average amplitude of the resulting voltage decreases as, more frequently, some area of target will be blocked by opaque spokes. Reticle B in Figure 6.6 can be used to determine the angle of the target based on when the detector starts to

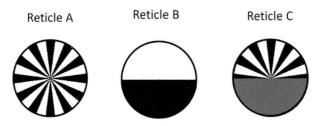

Figure 6.6   Reticle patterns.

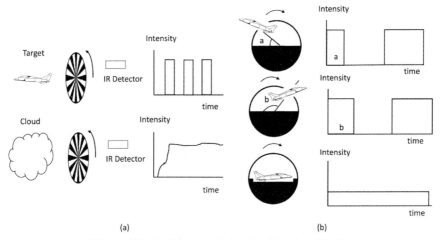

(a)   (b)

Figure 6.7   Working principles of reticles A and B [7].

detect the radiation. When the target is in the middle, the IR detector should detect a constant radiation. Figure 6.7 illustrated the working principles of these two reticles. Reticle C is a combination of reticles A and B; half of it is 50% transparent and half of it is a spokes reticle. So, it can determine object positions and differentiate targets from other background IR sources. The pattern of reticle C is referred to as a rising sun reticle and the goal of the missile guidance system using reticle C is to have the target locked in the center of the reticle. When the target is locked at the center of the reticle the IR detector generates constant small voltage. If not at the center, the target's position can be determined based on the amplitude of the pulse and the start time (phase) of pulses. This scan system is called a spin-scan system and used in Soviet Union's SA-7 IR SAM. As the spin-scan system directly looks at the aircraft, it is more vulnerable to jamming, and an IR conical scan system is developed to solve this issue.

The IR conical scan system uses the same principle as the conical scan radar covered in Section 2.13. The working principles of a conical scan system is illustrated in Figure 6.8. In an IR conical scan system, the second mirror swivels about the missile's roll axis and the target at the axis is always at the edge of rotating second mirror's field of view. This is done by tilting the second mirror and rotating it. The image reflected by the second mirror passes through a fixed full spoked reticle and the target at the center of the conical scan is mapped to the edge of the reticle. As the reticle is fixed and second mirror is rotating, the target at the center will "circle" around the reticle center, thus creating a periodic voltage change like a periodic pulse with a constant pulse width at the IR receiver's output. If the target is off the center, the pulse width will vary and this information can be used to correct the missile's direction. One might consider the conical scan system as a frequency modulation system that determines the target's location based on changing frequency (pulse width) and the spin-scan system as an amplitude modulation system that determines the target's position based on signal's amplitude and phase. As the field of view is not fixed with the target at the center but rotating around the target, the conical scan system covers a larger area. As a result, compared with a spin-scan system, a conical scan system has a smaller possibility of losing its target. The Soviet Union's SA-14 IR SAM uses the conical spin system to solve the jamming issues SA-7 encountered.

It is worthy of note that the IR detector does not provide range information. This information is not crucial as it is assumed that once the missile is adjacent to the target, the proximity fuze will detonate it. The reticle systems described in this section use only one detector. The energy of whole image is received by one IR detector and all tracking is based on this quantity's change over the time (a one-dimensional signal). In later developments, arrays of detectors (such as a focal plane array) are used to generate an image (a two-dimensional signal) for tracking.

6.7 Jamming Infrared Tracking System [9, 15]

Since the infrared tracking is passive, there is no guidance signal to detect. So, once an aircraft detects an approaching missile, the IR countermeasure must be applied immediately, even if the missile guidance system is not known.

As the IR missiles follow heat, their countermeasures are based on creating bright spots or directing radiation to the missile to deceive the missile's guidance system. The IR jammer can be either omnidirectional or directional. The omnidirectional jammer uses a high power infrared light source such

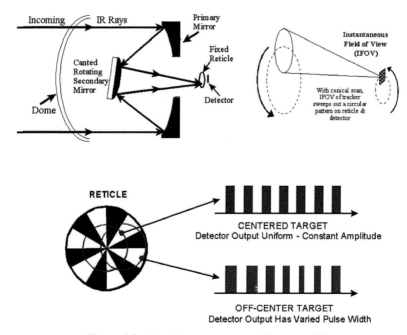

Figure 6.8   Conical scan system and its output [7].

as a heated silicon carbide block to radiate strong infrared energy and a shutter can be used to create a pulse pattern matching IR missile guidance system's reticle's pattern and spin speed. As the jammer's radiation is stronger than aircraft's radiation, hopefully, the missile guidance system will be confused by the jammer's radiation, thus losing its target. Once the missile's lock on the aircraft is broken, the missile has a slim chance of re-locking on its target. As different IR missiles might use different IR wavelengths, the jammer's light source needs to cover a broad spectrum to engage a variety of IR missiles. The effectiveness of this approach depends on whether the jamming radiation's wavelength and modulation pattern matches the IR missile's operation wavelength and its reticle's scan pattern. Moreover, the jammer's radiation energy needs to be sufficiently higher than aircraft's radiation energy (referred to as the jamming to signal (J/S) ratio) to be effective. This requirement is easier to achieve when jamming a spin-scan system because the IR missile looks directly at the aircraft. However, to jam a conical spin system, the required J/S ratio is much higher as the missile does not look at the aircraft directly.

To solve the jamming power issue of the omnidirectional jammer, directional infrared countermeasures (DIRCM) have been proposed and developed. DIRCM is integrated with a missile approach warning system previously described. After an approaching missile is identified and tracked, the DIRCM is turned on. Its IR source will be a multiple band infrared laser. Once the missile is tracked, this information can be used to aim the laser's output beam directly at the approaching missile. The laser's output will be modulated with different patterns in sequence so that it can circumvent a variety of IR missiles. As the DIRCM projects its energy directly toward the missile and the laser's beamwidth is narrow (unlike an omnidirectional light source whose energy radiates toward every direction), DIRCM can achieve a much higher J/S ratio.

Other effective countermeasure approaches against IR guided missiles are to create false targets by ejecting flares, which are high temperature burning object, or launching a decoy that mimics real aircraft. This subject will be discussed in the next chapter.

6.8 Conclusion

This chapter focuses on how to protect a target aircraft from an approaching missile and detecting the approaching missile is the first step. Some missiles do not need the guidance signal from a radar to lead an attack. In this case, a passive electronic warfare receiver designed to intercept radio signals cannot help.

Two missile detection approaches, an IR missile approach warning system which detects missile's IR radiation and a tail warning radar which detects the approaching missile using radio signals, are introduced. The IR warning receiver is a relatively simple approach. However, due to numerous heat sources in nature and reflections from the sun, the IR receivers suffer from a high false alarm rate. To reduce the number of false alarms and still detect the incoming missile is a very challenging technical subject. Multiple sensors covering different IR sub-spectra are used to distinguish the missile's exhaust plume/engine heat from the sunlight. These approaches complicate the receiver's design, but might reduce the system's false alarm rate. A complete radar system at the tail of an aircraft for detecting approaching objects is another way to find approaching missiles. This equipment is more suitable for large airplanes, such as transportation airplanes and bombers. The danger of such a system is that it can reveal an aircraft's position as it keeps emitting radio signals.

Unless a missile or an anti-aircraft artillery shell is designed to always hit its target aircraft directly, a proximity fuze is necessary. A brief history and the working principles of a proximity fuze are covered in this chapter along with methods designed to jam proximity fuzes.

Since a missile is a lethal weapon, once it is detected, jamming techniques must be applied immediately. In this chapter, IR missile guidance principles and possible jamming methods are explained. Just like jamming a radar, the intelligence about missile's guidance system is crucial for the jammer's success. However, this knowledge is not always available. If the approaching missile's guidance system is not known, all possible jamming techniques should be used. To jam an IR missile, both infrared jammers and flares (to be discussed in the next chapter) can be used. Obviously, the pilot may maneuver the aircraft to outsmart the missile (probably with the help of the sun), but this topic is beyond the scope of this book.

References

[1] Alfred Price, *History of US Electronic Warfare*, vol. 3, The Association of Old Crows, 2000.
[2] *Operation Desert Storm: Evaluation of Air Campaign*, United State General Accounting Office, June 1997.
[3] Anil K. Maini, *Handbook of Defence Electronics and Optronics: Fundamentals, Technologies, and Systems*, John Wiley & Sons, 2018
[4] J. F. Milthorpe, and P. J. P. Lynn, "Effect of Aerodynamic Heating on Infrared Guided Missiles," in: Z. Jiang (eds) *Shock Waves*, Springer, pp. 227–232, 2015.
[5] Joseph Trevithick, "US Army Hits Setbacks Trying to Add New Infrared Countermeasures to Its Helicopters," the WarZone, January 25, 2018. (https://www.thedrive.com/the-war-zone/17969/us-army-hits-setbacks-trying-to-add-new-infrared-countermeasures-to-its-helicopters).
[6] Robert L. Shaw, *Fighter Combat: Tactics and Maneuvering*, 7th edition Naval Institute Press, 1985.
[7] *Electronic Warfare and Radar Systems Engineering Handbook*, Naval Air Warfare Center Weapons Division, 2013.
[8] Sungho Kim, Sun-Gu Sun, and Kyung-Tae Kim, "Analysis of Infrared Signature Variation and Robust Filter-Based Supersonic Target Detection," Scientific World Journal, vol. 2014, Article ID 140930, 2014.
[9] G. Kim, B. Kim, T. Bae, Y. Kim, S. Ahn, and K. Sohng, "Implementation of a Reticle Seeker Missile Simulator for Jamming Effect Analysis,"

in the Proc. of the 2nd International Conference on Image Processing Theory, Tools and Applications, Paris, 2010, pp. 539–542.

[10] Edward A. Sharpe, "The Radio Proximity Fuze - A Survey," *Vintage Electrics*, vol. 2, no. 1, https://www.smecc.org/radio_proxi mity_fuzes.htm

[11] Richard E. Marinaccio and Ward M. Meier, "Proximity fuze jammer," US Patent # US4121214A, 1969.

[12] John W. Lyons. E. A. Brown, and B. Fonoroff, "Radio Proximity Fuzes," https://nvlpubs.nist.gov/nistpubs/sp958-lide/059-062.pdf.

[13] W. S. Hinman and C. Brunetti, "Radio Proximity-Fuze Development," in Proceedings of the IRE, vol. 34, no. 12, pp. 976–986, Dec. 1946.

[14] Graham Warwick, "Blinding with Science," Flight International, 10 January 2000 (https://www.flightglobal.com/blinding-with-science/ 30120.article)

[15] C. Kopp, "Heat-Seeking Missile Guidance", Australian Aviation, 1982

[16] Tae-Wuk Bae, Byoung-Ik Kim, Young-Choon Kim, and Sang-Ho Ahn, "Jamming Effect Analysis of Infrared Reticle Seeker for Directed Infrared Countermeasures," Infrared Physics and Technology, vol. 55, no. 5, pp. 431–441, Sep. 2012.

[17] T. M. Malatji, W. P. du Plessis, and C. J. Willers, "Analysis of the Jam Signal Effect against the Conical-scan Seeker," Optical Engineering, vol. 58, no. 2, Article ID 025101, 2019.

[18] C. J. Willers, J. P. Delport, and Maria Susanna Willers, "CSIR Optronic Scene Simulator Finds Real Application in Self-protection Mechanisms of the South African Air Force," in the Proc. of Science Real and Relevant Conference, 2010.

[19] Sam Goldberg, "Infrared Countermeasures: the Systems That Cool the Threat from Heat-seeking Missiles," Air and Space Magazine, July 2003. (https://www.airspacemag.com/how-things-work/infrared- countermeasures-4739633/).

[20] J. Heikell, *Electronic Warfare Self-protection of Battlefield Helicopters: a Holistic View*, Helsinki University of Technology, Applied Electronics Laboratory Series E: Electronics Publications E18, Espoo-Finland, 2005

[21] Jack R. White, *Aircraft Infrared Principles, Signatures, Threats, and Countermeasures*, Naval Air Warfare Center Weapons Division, Sep. 2012.

[22] *Electronic Warfare Fundamentals*, U.S. Department of Defense, November 2000.

7

Chaff, Flare, and Decoys

7.1 Introduction

We have gone through the fundamental radar and electronic warfare principles. The major purpose of jamming is to prevent a radar or a missile guidance system from obtaining accurate location/speed information about its target. This task can be accomplished by emitting radio waves (against a radar) or IR radiation (against an IR missile) as described in previous chapters. In this chapter, several countermeasures not based on electronic devices will be introduced. Instead of sending out radio waves or infrared light, the approaches covered in this chapter use physical objects to produce fake targets on the radar screen or heat sources to attract IR missiles. Although some of these countermeasures have a history just slightly shorter than radar's, they are still in use today, and we will start with the oldest one, chaff.

7.2 Chaff and its Countermeasures

Among all jamming techniques, chaff is the oldest one, but it is still in use. In the early days, chaff was a strip of metal coil, but, nowadays, chaff is made of nylon or fiberglass coated with metal and a cartridge can hold an enormous amount of chaff. The working principle of chaff is simple. As a radar sends out radio waves and determines its target's location based on the reflected signal, if many objects reflecting the radar signal back to the radar are dispensed at once, they can create a cloud of false targets on the radar screen and the radar operator will have a difficult time figuring out how to respond. Chaff is designed to reflect radar signal, thus creating such confusion. As the chaff is resonant at a wavelength that is twice its length (in other words, the chaff length should be half of its victim radar signal's wavelength), chaff of several carefully chosen lengths can be dispensed together to ensure maximum frequency coverage and reflection. The whole

113

purpose of chaff is to create a false target image and different methods of dispensing chaff have been developed to fulfill this goal. For example, a large quantity of chaff can be dropped for a period of time by a supporting aircraft to create a chaff corridor to mask following aircraft. This practice is called stream chaff. How fast chaff should be dispensed depends on the size of the radar's resolution cell (the minimum volume of space for a radar to distinguish its targets). Too much chaff dispensed in a small space does not help and this practice can be wasteful. On the other hand, too little chaff dumped within a resolution cell might not be enough to create reflected radar signals strong enough to be detected. As these chaff dispensing techniques need to be practiced, chaff's environmental effects raised concerns. Some studies on this matter have been conducted and found no significant adverse impact on environment and human/animal health.

Once the chaff is dispensed, it deccelerates quickly so that a Doppler radar that can detect its target's velocity is able to differentiate between chaff and its target based on their speed. In this case, the pilot might fly in a tangential direction to the radar beam so that the aircraft's speed measured based on Doppler shift is close to zero and the chaff can offer protection.

7.3 Battle Example: Bombing of Hamburg on July 24, 1943

On the evening of July 24, 1943, German radars facing the North Sea detected thousands of approaching aircraft that were able to "reproduce" themselves. German's night fighters who were called to engage the incoming bombers followed the instructions of radar operators but could not find enemy aircraft. Later, every radar set in Hamburg was out of action as the whole radar screen was clouded by flickers, thus not being able to provide any useful guidance to anti-aircraft battery commanders. So, radar operators reported the malfunction of their radars. Despite German radar detecting thousands of fake targets, 791 British bombers did fly to raid Hamburg. Without an effective German air defense, British pilots completed their mission pretty much undisturbed and lost only 12 bombers. The loss rate of this mission was 1.5% and the loss rate in the previous six raids on Hamburg was 6.1%.

What German radars detected on that night was tin-foil strips, i.e. chaff, dropped by the British Royal Air Force. The British code name for chaff is "window," and the codewords "open the window" mean dispensing chaff. The bombing of Hamburg on that night was the first successful usage of chaff in battle. One interesting fact about chaff is that its basic concept is so simple that both British and German scientists thought about it before 1943 (a British cartoonist, Jack Monk, came up with the idea of Nazi's using metal-framed box-kites to confuse British radar and published a cartoon strip on this idea in the *Daily Mirror*, thus causing serious concern among British officials involved with the "window" project). The German code name for chaff is Düppel. After learning of the jamming effects of Düppel, Hermann Göring ordered its development to stop immediately and destroyed related research documents, fearing such a simple technology would fall into the enemy's hands. The British government also hesitated to introduce "window" for the same reason and the final approval of using chaff was given by Winston Churchill who refused to discuss this matter in the war cabinet as it was too technical.

After a while, German radar operators developed techniques for distinguishing between chaff and real targets, including using speed to separate them and changing radar frequencies if chaff was present. German also used Düppel in their retaliation.

7.4 Flare and its Countermeasures

The Oxford Learner's Dictionary's definition of flare is "*a bright but unsteady light or flame that does not last long.*" It is also a pretty accurate description of flare used in electronic warfare as, in this context, flare is a very hot object ejected from an aircraft and its heat or brightness lasts for a short period of time (several seconds). Just like chaff is designed to appear as real targets for radar, flare is designed to generate IR radiation fooling an IR missile's guidance systems into mistaking it for a true target. To achieve this goal, flare must (1) reach peak intensity almost immediately after being ejected, (2) create an IR radiation stronger than the radiation emitted by the aircraft's heat sources like the exhaust plume and engines, and (3) possess an IR spectrum more attractive than the IR spectrum of aircraft the IR missile goes after. Flare can be made of either pyrotechnic material which generates highly visible white light or pyrophoric materials which ignite immediately after exposed to oxygen, thus generating IR radiation.

In an ideal scenario (for the target aircraft, of course), once a flare is ejected, the IR missile's heat seeker begins to track the flame and lose its lock on the aircraft. Flare can be a very effective countermeasure against IR missiles using a spin-scan system. Just like chaff, flare starts to lose its speed immediately after ejection. To distinguish between the flare and the real target, the IR missile guidance can use speed to separate the flare from the real target like a radar using speed to distinguish between aircraft and chaff.

For a conical scan system which rotates around its target, the flare might fall out of the conical scan system's field of view due to it losing speed too soon for the conical scan system to track it. The IR missile can also detect flare by noticing a sudden increase in IR radiation via comparing the IR spectra of flare and target or by sudden separation of two IR sources (i.e. flare and the aircraft). When a flare is detected, the missile's tracking system might wait until the flare is out of its field of view to track its target again.

To counter the IR missile guidance system's counter-countermeasure, flare should have an IR spectrum close to the one of the aircraft it is designed to protect, and multiple flares can be ejected continuously as the IR missile's tracking accuracy is sacrificed if the flare stays longer in its field of view. Advanced flare is even designed to propel itself behind the aircraft for a while, so the IR missile cannot identify a flare by its sudden separation from the aircraft or its deceleration. Such a flare is called a kinematic flare. One can argue that a flare with a mechanism to propel itself is similar to a glider or decoy, and it will be the subject of the next section.

7.5 Decoy

Like chaff, a decoy is built to confuse radar so that radar will mistake it for a target aircraft. The major difference is that a decoy is way more complicated. It can be considered as a miniature aircraft and some decoys are powered and carry a jammer. To maximize the chance of deceiving a radar, a decoy is designed to increase its radar cross section so that a small decoy can appear as a big jet fighter or bomber on a radar screen.

A decoy can be very useful in several ways. Based on their usage, they might be classified as expendable decoy, towed decoy, or saturation decoy. The expendable decoy is ejected from the aircraft to distract the missile. The electronics on an expendable decoy can generate large radar images to lure missiles, thus protecting the aircraft. Advanced expendable decoys such as the BriteCloud series are equipped with DRFM to capture and retransmit the radar signal to lure missiles from the aircraft. One issue of expendable decoys is that, although an expendable decoy's electronics can fool a radar, just like chaff, expendable decoys do not have an engine, so it is like a glider which cannot stay in the air for an extended time. This issue can be solved by a towed decoy. From its name, one can image that a towed decoy is attached to an aircraft so that it can fly behind. The modern towed decoy is connected to the aircraft by an optical cable and it can transmit the jamming signal generated by the jammer of the aircraft. If jamming does not work, the last resort will be letting the towed decoy be the target of the missile. After a successful mission, if not destroyed, the towed decoy can be reeled in for reuse. One example of a towed decoy is the AN/ALE-55 fiber optic towed decoy used in F-18s.

The main function of a saturation decoy is different from the two decoys just described. Instead of protecting aircraft, a saturation decoy is used to trigger the enemy's radars into action. To fulfill this purpose, new saturation decoys such as the US Air Force's ADM-160B Miniature Air-Launched Decoys (MALD) are equipped with engines so that they can fly for an extended distance (several hundred miles) and have a long air time (45 minutes). Such a decoy is also referred to as a decoy missile. As saturation decoy possesses a radar cross section and flight characteristics similar to real aircraft; it can lead the first wave of attack to bait the enemy's radars into operation, thus wasting the enemy's resources for tracking real attacking aircraft and exposing the enemy's radar locations. The function of saturation decoys can also be fulfilled by drones. As described in Section 5.14, in the 1982 Lebanon War, the Israeli Air Force used drones to trigger Syria's radars into operation so that the Israeli Air Force could precisely locate and neutralize 17 out of 19 Syrian SA-6 sites in 15 minutes. On the first night of Operation Desert Storm, drones and the US Navy's unpowered TALD (Tactical Air Launched Decoy)[1] were used to trick Iraqi anti-aircraft guns, missiles, and radars into action. Many of them were then destroyed and the Iraqi Air Defense was significantly reduced from the beginning of the first Gulf War.

7.6 Conclusion

In this chapter, three non-electronic countermeasures, chaff, flare, and decoy, are introduced. Up to now, we have described the cycle of electronic warfare as follows: a radar sends out radio wave to detect targets and an IR missile using IR radiation from a target to track it. The countermeasure is developed to create confusion for radar or IR missile's tracking system and counter-countermeasures are then conceived to reject countermeasures accordingly. Every countermeasure will meet with its counter-countermeasure, and every counter-countermeasure will trigger the development of more advanced countermeasures. It is the circle of electronic warfare that never ends. Another way

---

[1]Like MALD, TALD have both powered and unpowered versions.

to win this competition can be by preventing it from starting. If an aircraft can be designed in such way that it cannot be seen by a radar in the first place, then there is no need to deceive radar with electronic and/or non-electronic jamming methods, thus avoiding all of the trouble afterwards. Such aircraft is often referred to as stealth aircraft which will be the topic of the next chapter.

References

[1] George W. Stimson, *Introduction to Airborne Radar*, 2nd edition, SciTech Publishing, 1998.

[2] Anil K. Maini, *Handbook of Defence Electronics and Optronics: Fundamentals, Technologies, and Systems*, John Wiley & Sons, 2018.

[3] *Electronic Warfare and Radar Systems Engineering Handbook*, Naval Air Warfare Center Weapons Division, 2013.

[4] *Electronic Warfare Fundamentals*, U.S. Department of Defense, November 2000.

[5] Richard E. Farrell and Steven D. Siciliano, *Effects of Radio Frequency (RF) Chaff Released during Military Training Exercises: A Review of the Literature*, (https://www.gov.nl.ca/ecc/files/env-assessment-projec ts-y2004-1159-environmental-effects-of-radio-frequency-chaff.pdf), March 2004.

[6] R. V. Jones, *Most Secret War*, Wordsworth, 1978.

[7] Alfred Price, *Instruments of Darkness: the History of Electronic Warfare 1939–1945*, Frontline Books, 2017.

[8] Mario de Arcangelis, *Electronic Warfare: From the Battle of Tsushima to the Falklands and Lebanon Conflicts*, Blandford Press, 1985.

[9] Grant Turnbull, "Can New Spoofing Tech Give US Aircraft a Shroud in the Clouds?," Defense News, (https://www.defensenews.com/electronic -warfare/2019/05/15/can-new-spoofing-tech-give-us-aircraft-a-shroud -in-the-clouds/), May 15, 2019

[10] John Keller, "Navy asks BAE Systems to Build T-1687/ALE-70(V) Electronic Warfare (EW) Towed Decoys for F-35," Military and Aerospace Electronics, (https://www.militaryaerospace.com/unma nned/article/16726515/navy-asks-bae-systems-to-build-t1687ale70v-e lectronic-warfare-ew-towed-decoys-for-f35), Aug. 17, 2018.

[11] Tyler Rogoway, "Recent MALD-X Advanced Air Launched Decoy Test Is A Much Bigger Deal Than It Sounds Like," the Warzone, (https://www.thedrive.com/the-war-zone/23126/recent-mald-x-ad

vanced-air-launched-decoy-test-is-a-much-bigger-deal-than-it-sounds-like), Aug. 24, 2018.

[12] Carlo Kopp, "Operation Desert Storm: The Electronic Battle Parts 1–3," *Australian Aviation* (http://www.ausairpower.net/Analysis-ODS-EW.html), June/July/August, 1993.

[13] Eliot A. Cohen, *Gulf War Air Power Survey: Gulf War Air Power Survey: Weapons, Tactics, and Training and Space Operations*, Office of the Secretary of the Air Force,1993.

8

Stealth Aircraft

8.1 Introduction

One of the unconfirmed stories about stealth aircraft goes as follows. Once upon a time, a radar operator was told to do a field test for a radar, and he duly performed his duty. While he was working on testing his radar, he heard a roaring sound of aircraft jet engines. Looking toward the direction the engine noise was from, he spotted an approaching aircraft. However, when he checked his radar screen, he found nothing. So, he began to make adjustments, but none of the changes he made worked, and up until the aircraft disappeared over the horizon, his radar still could not find this aircraft. This radar operator began to think he was given a defective radar. "What a lucky day!" he told himself and then got ready to take the radar back for repair. What he did not realize was that he actually successfully finished a test that was not meant for the radar but for the stealth aircraft he saw.

Chi-Hao Cheng heard this story from one of his colleagues but could not verify it. So, this story might well be another urban myth. Nevertheless, it does capture what a stealth aircraft is designed for. It is designed to not be detected by a radar, especially a fire control radar. Jamming techniques for radars are needed because radars can detect aircraft and aircraft need protection from being locked onto by hostile radars. So, in principle, a stealth aircraft does not need to jam. Of course, as we will see, a stealth aircraft is not an invisible aircraft. In the following sections, we will introduce the fundamental principles of stealth aircraft and how they might be detected.

8.2 Stealth Technologies

To avoid being detected by a radar, one straightforward way of thinking is covering the aircraft with specially engineered Radar-Absorbing Material (RAM) to reduce the energy of the reflected radar signals. The US's F-117 uses iron ball paint absorber and the F-35 uses "fiber-mat" materials. Other materials include foam absorber, Jaumann absorber, etc. The composition and usage of RAM can be classified and kept as a well-guarded secret. When Lockheed Martin declared its use of a fiber mat for the F-35, it declined to provide details about this technology as it is classified. In 1999, an F-117 was shot down in Yugoslavia and it was reported that some of wreckage was sent to China and Russia to help develop their stealth technology. When asked about Russia and China possibly gaining knowledge about US stealth technology from the F-117 wreckage, US Maj. General Bruce Carlson responded "I would just tell you this: The science involved in making a low-observable airplane is not a secret. It involves shaping and radar absorbing material. The technology needed to make radar absorbing material is available in a number of places." [3]

General Carlson probably spoke the truth. After all, some RAM like (CNT)-infused fiber material has been granted a US patent (patent number US20100271253A1) and everyone can search for these patents online. However, avoiding a radar's detection is one thing and being used in a combat environment is another. The coating of a jet fighter should be light, strong, durable, and easy to repair. But some RAM like iron ball paint can be costly, heavy, and difficult to maintain. After each mission, the stealth aircraft surface needs to be carefully inspected. One of James Tsui's friends who used to work on stealthy technology told him that stealth aircraft need to be frequently repainted. It is not difficult to image how expensive and inconvenient this practice can be. Nevertheless, this issue might be solved now as Lockheed Martin has claimed that its fiber-based stealth technology is a "durable, low-maintenance stealth technology."

Compared with RAM, another way to avoid radar detection is much less straightforward. For a very longtime, the common wisdom was that the bigger the aircraft, the bigger its radar cross section. Therefore, a big aircraft should appear as a large object on a radar screen and a decoy is designed to mimic an aircraft's large radar cross section. In 1962, a Soviet Union physicist, Pyotr Ufimtsev, published his study on the scattering of electromagnetic waves and this work is known as the Physical Theory of Diffraction. In this work, Ufimtsev built a foundation for scattering incident radar electromagnetic waves by carefully designing abrupt surface discontinuities or sharp edges so that signals reflected back to radar can be minimized [4, 5]. In other words, a carefully designed aircraft can have a very small radar cross section. Although the Soviet Union did not consider this approach practical, thus not classifying Ufimtsev's work, the US recognized its importance and began to test this concept. Ufimtsev's work was translated into English and can be freely downloaded from the Internet now [5].

One issue about stealth aircraft is that it adds additional constraints to the aircraft design. As we have seen in the case of RAM, to design an aircraft that can scatter radar signals so that radar cannot detect the aircraft is one thing; to have an aircraft frame with controllability and maneuverability suitable for an air battle is another. When the US began to test Ufimtsev's theory, Lockheed Martin called its prototype the Hopeless Diamond, which is shown in Figure 8.1. This name is a pun on the famous Hope Diamond. One can see from the figure that the "diamond" part of this nickname is a reference to the shape of aircraft. The "hopeless" part is a tribute to the design challenge accompanying the stealth requirement. Even after making stealth aircraft for more than four decades, in 2018, Lockheed Martin's management stated that half of the F-35 defects resulted from its stealth features. To some degree, designing a stealth aircraft is like designing a silent race car to win an F-1 race. Of course, the philosophy of stealth aircraft is that if the enemy's radar cannot detect the aircraft, then the fight is over before it even starts as the stealth aircraft can conveniently destroy its opponents before they are aware of its existence. It is like the race car driver driving a silent race car starting and finishing the race before other race car drivers realize that the race has started.

Figure 8.1 Hopeless diamond (https://en.wikipedia.org/wiki/Lockheed_Have_Blue#/media/File:Have_Blue_bottom_view.jpg).

Besides reducing the probability of an aircraft being detected by a radar, a stealth aircraft should also reduce its heat emission, visibility to the naked eyes, etc. As this book is mainly about radar and EW systems, these topics are not covered here.

8.3 Passive Radar for Stealth Aircraft?

The stealth aircraft is designed to be invisible to many radars, but it might not be so invisible if illuminated by a radio wave of lower frequency such as broadcasting TV or radio signals whose frequency is lower than the frequency of the radars that the stealth technology is designed against. The reason is that a lower frequency radio signal has a longer wavelength (wavelength = speed of light divided by frequency) and for a long-wavelength radio signal, the whole aircraft is like an antenna reflecting signals back. A passive radar does not emit signals to actively search for targets. Instead, it relies on signals emitted by other sources such as radio and TV stations and signals reflected from the aircraft. By comparing signals transmitted directly from a transmitter at a known location, say a TV tower, and signals reflected from an object, the passive radar can locate its object. In 2019, German radar maker Hensoldt claimed to have used its passive radar to track two F-35s attending the 2018 Berlin Air Show in Germany for 150 km and this news stirred up quite a few discussions [7].

So, what is the issue of using a radar with longer wavelength to track stealth aircraft? The answer is the precision. A longer wavelength radar has a coarse resolution. Simply knowing an aircraft is in a ballpark range is not good enough to guide a missile or anti-aircraft artillery against it. Hensoldt also admitted that their radar precision was not enough to guide a missile (yet). On the other hand, just like a commercial GPS receiver uses signals from multiple satellites and a network of fixed reference stations to improve its precision, it is also possible to use a network of passive radars exploiting multiple transmitters to improve the precision. The game between radars and stealth aircraft surely will continue.

8.4 Battle Example: 1999 F-117A Shootdown

On March 27, 1999, three days after NATO began bombing Yugoslavia to force Yugoslav (mainly Serbian) armed forces to withdraw from Kosovo, the Yugoslav army achieved something no other country had ever managed to accomplish before or since: shooting down a stealth fighter. More astoundingly, the Yugoslav army used the Soviet Union's SA-3 surface-to-air missile (SAM) designed in the 1960s to bring down US's F-177 which was successfully used in the 1990–1991 Operation Desert Storm. Soon, the Serbian government made anti-NATO posters with the catch phrase "Sorry, we didn't know it was invisible" (see Figure 8.2). So, how could it be possible? The official report on this episode might remain inaccessible to public for a long time, but an explanation based on a US Air Force aerospace engineer Chris Morehouse's online post is pretty plausible.

Aware of the US's EW capability, the Serbian force kept relocating its missile sets; therefore, unlike the Iraqi force losing its anti-aircraft capability at the beginning of Desert Storm, Serbian's SAMs were largely intact after the bombing started, and they could set up a missile set quickly to attack. Probably, due to the US's overconfidence about its air superiority, US's F-117 were set on the same route to strike and this practice gave Serbian force an opportunity to prepare for its ambush along the route. The Serbian SA-3

Figure 8.2 Serbian poster (https://www.reddit.com/r/dragonutopia/comments/d6fom6/ sorry_we_didnt_know_it_was_invisible_propaganda/).

system used two radars: a P-18 early warning radar and an SNR-125 fire control radar. The frequency of the P-18 is VHF (30–300 MHz). Part of the VHF spectrum is used for FM and TV broadcasting in the US. Serbian force found that if they used the lowest working frequency of the P-18, they could detect F-117s (just like a passive radar might detect a stealth aircraft by exploiting FM/TV signals). However, low frequency radar does not provide good precision and the P-18's range was only 15 miles when operated in this setting. Nevertheless, it was good enough to provide warning if the flight route was expected.

On March 27, 1999, the US's electronic warfare aircraft were grounded due to weather conditions and Serbia was aware of this situation, thanks to its informers. Therefore, when F-117 flew close by on that day, Serbia's SNR-125 fire control radar could make several tries without fear of being tracked and targeted. During one of the tries, one F-117 was preparing to drop its ordnance and opened its weapon bay. As discussed in Section 8.2, the stealth aircraft's shape is carefully designed to reduce its radar cross section and its body is covered by RAM to absorb radar signals. When the weapon bay is opened, the aircraft no longer maintains its magic shape and its interior reflects radar signals. As a result, the SNR-125 was able to lock on the F-117, firing two SA-3 missiles toward it and one brought down the invisible F-117. The F-117 pilot, Lt. Col. Dale Zelko, was ejected and later rescued.

In warfare, the element of surprise is crucial. One of the major reasons for the US's F-117 and Taiwan's U-2 being shot down (Section 3.9) is that their flight routes were correctly predicted. Neither of them had appropriate electronic support when they were shot down. Of course, a stealth aircraft is not supposed to be locked onto by a 1960s fire control radar under a normal circumstance; so it was Zelko's bad luck for being brought down when his aircraft was in the most vulnerable (or least invisible) mode. Nevertheless, this embarrassing episode of US military action could have been avoided.

8.5 Conclusion

Stealth aircraft were developed based on an idea that if an aircraft cannot be detected by a radar, then it is immune to the radar's threats. The same idea can also be applied to radar. If a radar can send probing signals without being detected by an intercept receiver, then it will not be jammed unless the jammer is willing to expose itself by emitting jamming signals when it does not detect a radar in operation. Even in this case, the jamming might not be effective as the jammer has no information about radar signal's characteristics such as frequency. A radar designed to not be detected is called a low probability of intercept (LPI) radar which will be the subject of the next chapter.

References

[1] Bahman Zohuri, *Radar Energy Warfare and the Challenges of Stealth Technology*, Springer 2020.

[2] Tom Hundley, "Serbs Sell Secrets of Downed Fighter," Chicago Tribute, November 22, 2001.

[3] Relly Victoria Petrescu and Florian Ion Petrescu, *Lockheed Martin Color*, Florian Ion Petrescu, 2013.

[4] Konstantinos Zikidis, Alexios Skondras, and Charisios Tokas, "Low Observable Principles, Stealth Aircraft and Anti-Stealth Technologies," Journal of Computations & Modelling, vol. 4, no. 1, pp. 129–165, Jan. 2014.

[5] P. Ya. Ufimtsev, *Method of Edge Waves in the Physical Theory of Diffraction*, 1962 (https://apps.dtic.mil/sti/pdfs/AD0733203.pdf).

[6] Valerie Insinna, "Stealth Features Responsible for Half of F-35 Defects, Lockheed Program Head States," Defense News, March 5, 2018 (https://www.defensenews.com/air/2018/03/06/stealth-features-responsible-for-half-of-f-35-defects-lockheed-program-head-states/).

[7] Sebastian Prenger, "Stealthy No More? A German Radar Vendor Says it Tracked the F-35 Jet in 2018 – from a Pony Farm," September 29, 2019 (https://www.c4isrnet.com/intel-geoint/sensors/2019/09/30/stealthy-no-more-a-german-radar-vendor-says-it-tracked-the-f-35-jet-in-2018-from-a-pony-farm/).

[8] Dan Katz, "Stealth - Part 2 - Physics And Progress Of Low-Frequency Counterstealth Technology," Aviation Week & Space Technology, August 25, 2016.

[9] W. Lei, W. Jun and X. Long, "Passive Location and Precision Analysis Based on Multiple CDMA Base Stations," in the Proc. of 2009 IET International Radar Conference, Guilin, 2009, pp. 1–4.

[10] Dario Leone, "An-in-Depth Analysisof How Serbs Were Able to Shoot Downan F-117 Stealth Fighterduring Operation Allied Force," The Aviation Geek Club, March 26, 2020 (https://theaviationgeekclub.com/an-in-depth-analysis-of-how-serbs-were-able-to-shoot-down-an-f-117-stealth-fighter-during-operation-allied-force/).

9

Low Probability of Intercept Radar

9.1 Introduction

In the field of electronic warfare, radar is designed to detect a signal, which it emits and then is reflected back by aircraft. So, the radar knows which signal to look for. On the other hand, the electronic warfare system might not have *a priori* knowledge about the radar signal reaching the aircraft, so it needs to search for signals over a broad spectrum and in every direction. However, the EW system's intercept receiver detects signals from the radar and the radar is designed to detect signals reflected by the aircraft that make a round-trip between the radar and target aircraft. So, in theory, the intercept receiver can detect a radar signal before the radar receives the reflected signal and, if we assume that the intercept receiver and the radar receiver have the same sensitivity, the electronic warfare system can identify the existence of the radar before the radar finds the aircraft. Under these circumstances, the radar is in a disadvantaged position. As we have seen in the example of the 1982 Lebanon War and Desert Storm, one of the strategies to gain air superiority is to trick the enemy's radars into searching for decoys and to destroy enemy radars by following radar signals.

To compensate for this disadvantage, some radars are designed to be difficult to intercept and such radars are referred to as low probability of intercept (LPI) radar which is the topic of this chapter. Before covering the LPI radar's approach, some re-examination of radar from the LPI radar's perspective can be beneficial.

9.2 Review of Radar Concepts

Up to this point, radars considered in this book are mainly pulse radars. A pulse radar sends out pulses periodically. Based on when the received signal is received, the target can be located. The pulse width (PW) needs to be short to guarantee a good distance resolution (Equation (2.3)) and the energy of transmitted pulse needs to be high enough to guarantee the detection of signal reflected back within the radar's working range. As a result, the power of the pulse needs to be high as power is calculated by dividing the energy by the PW. Since the pulse is sent out periodically, if the reflected pulse is received after the next pulse is emitted, the target cannot be accurately located. Therefore, the pulse repetition interval determines the radar's working range. The reflected signal's frequency might be altered because of the Doppler effect and this information can be used to determine the target's speed. To search for its target, a radar needs to keep scanning through the sky and some scanning patterns were introduced in Chapters 2.12–2.14. The radar's antenna has mainlobes and sidelobes, as shown in Figure 2.7, and the radar's antenna transmits and receives signals through both mainlobe and sidelobe although when the radar detects a signal, the radar assumes the signal is through the mainlobe as the antenna's mainlobe gain is much higher than its sidelobe gain.

As we keep mentioning in this book, in principle, one major advantage the EW system holds over radar is that the intercept receiver intercepts the signal directly from the radar. In Chapter 2, an equation relating the radar's transmitted signal power and received signal power is given and it is restated below for reference

$$P_{\text{rec}} = P_t G \frac{\sigma}{4\pi r^2} \frac{\frac{G}{4\pi}\left(\frac{c}{f}\right)^2}{4\pi r^2} = P_t G^2 \frac{\sigma}{(4\pi)^3 r^4} \left(\frac{c}{f}\right)^2 \tag{9.1}$$

where P_{rec} is the power of the received signal, P_t is the power of the transmitted signal, σ is the target's cross section area (i.e. the area that reflects radar signals, which the stealth aircraft aims to reduce), G is the radar's antenna gain, the term $\frac{G}{4\pi}\left(\frac{c}{f}\right)^2$ is the receive antenna's effective area, c is the speed of light, f is the signal frequency, and r is the distance between the radar and its target. For an intercept receiver, the relation between its intercepted signal power and transmitted signal power can be written as

$$\tilde{P}_{\text{rec}} = P_t G \frac{1}{4\pi r^2} \frac{\tilde{G}}{4\pi} \left(\frac{c}{f}\right)^2 \tag{9.2}$$

where $\tilde{P}_{\text{rec}}$ is the power of the intercepted signal, $\tilde{G}$ is the intercept receiver's antenna gain, and the term $\frac{\tilde{G}}{4\pi}\left(\frac{c}{f}\right)^2$ is the intercept receiver's antenna effective area. So, for a target whose distance from the radar is r, the ratio between the EW system and radar received signal power can be obtained by dividing Equation (9.2) by Equation (9.1), and we get

$$\frac{\tilde{P}_{\text{rec}}}{P_{\text{rec}}} = \frac{\tilde{G}}{G\sigma} 4\pi r^2 \qquad (9.3)$$

So, if every parameter in Equation (9.3) were fixed except for r, if we increase r, at some point, the EW system's received signal power will be higher than the power of the signal reflected back to the radar. As a consequence, an EW system can detect the existence of radar before the radar detects the aircraft beyond this distance, **provided that the EW system knows where and how to look for the radar signal**. The catch is, of course, the EW system needs to know what it is looking for and the LPI radar is designed in such way that the EW system might miss the radar signal in front of its face until it is too late.

9.3 Overview of LPI Radar Technologies

The main goal of LPI radar is to avoid being detected or at least detect the aircraft before being detected by the intercept receiver on the aircraft. To accomplish this goal, many approaches can be used. In this section, some of LPI radar technologies are introduced and two subsequent sections will focus on two specific LPI radars.

Minimize Antenna Sidelobe. When a radar emits signals, the signals are intended to be sent through antenna's mainlobe and the signal transmitted through the antenna's sidelobe can be considered leakage. The signals emitted through the sidelobe not only expose the radar, but the reflected radar signal through the sidelobe, if detected, confuses the radar as it assumes the received signal is via the antenna's mainlobe. As a result, the EW system can jam a radar by aiming the jamming signal at the radar antenna's sidelobe. Therefore, an LPI radar should have a small antenna sidelobe and very low sidelobe gain.

Use Complicated Search Patterns. A radar scans through the sky looking for its target following a certain pattern. Once this pattern is recognized, the corresponding jamming method can be designed. One example is the inverse gain jamming covered in Section 5.9. The more complicated the scan pattern is used, the more difficult it is for the EW system to deceive the radar. With the advance of phased array radar, which uses a matrix of stationary antennas to electronically steer the radar signal to a desired direction by changing the phase of each antenna's transmitted signal, it is much easier to program a sophisticated scan pattern. It is worthy of note that phased array radar can also better control its collective sidelobe.

Frequency Hopping. The radar can reduce the chance of being detected by changing its frequency frequently even on a pulse-to-pulse basis. Such a practice can complicate the EW system's effort to sort radar signals.

Power Management. To reduce the chance of being intercepted, the radar signal's power should not be higher than necessary. Therefore, once a target is found, based on the power of reflected radar signal, the radar might reduce the power of the transmitted signal to just enough to keep tracking the target.

Pulse Compression. The techniques for LPI radar covered so far do not address the fundamental issues summarized in Equations (9.1)–(9.3), i.e. the intercept receiver can intercept a radar's signal far away because the reflected radar signal travels double the distance, thus suffering higher loss when it reaches the radar. However, the radar possesses one huge advantage over the EW system: the radar knows its own signal. So, one of the key LPI radar's design concepts is to reduce the transmitted signal power by spreading the signal energy over a longer time; thus, the radar signal's power will be too low to alarm the intercept receiver. The radar can integrate the reflected signal's power over a longer time, thus obtaining enough energy to detect the target. On the other hand, the intercept receiver will likely miss the radar signal as it does not know how long it should keep "listening" to detect the incoming signal and it might not have information about signal characteristics to retrieve the signal. If the intercept receiver only integrates the received signal's energy over a short period, the resulting output can be too weak to be detected despite the fact that the power of the intercepted radar signal can be significantly higher than the power of reflected signal that the radar receives. Besides, simply integrating the received signal does not provide enough information about how to jam a radar. An analogy of this situation can be a spy trying to eavesdrop a conversation between two people who whisper an encrypted message while playing loud music. The voice volume is very low, so the spy might not notice the ongoing conversation and knowing there is a conversation alone is not enough to steal the secret.

Of course, one immediate concern about this approach is that a long pulse can sacrifice distance resolution which equals the speed of light times pulse width of a pulse radar divided by 2 (Equation 2.3). Also, for a pulse radar, once a pulse is sent, the radar will be silent for some time waiting for the reflected signal before the next pulse is sent. For example, a radar can have 1 ms of PRI and 1 µs of PW. In this case, in every millisecond, the radar only transmits a signal for 1 μs (1/1000 of 1 ms) and wait for reflected signal for 999 μs (999/1000 of 1 ms). If a long pulse is used, then the radar needs to wait for an even longer time to receive the signal reflected back from the aircraft. Consequently, the radar's response time can be too long. To solve these issues, pulse compression techniques need to be applied. In the following two sections, two popular LPI radars based on pulse compression principles, phase coded radar, and frequency modulated continuous wave (FMCW) radar will be explained.

9.4 LPI Radar Example 1: Phase Coded Radar

Before introducing two LPI radar examples, the concept of cross-correlation needs to be explained first. Cross-correlation is used to measure similarity between signals, and we will use radar signals as an example to explain how it is calculated. Assume a radar sends out one cycle of a sinusoidal wave whose period is 20 ns (PW = 20 ns) and wait for its reflection. The top plot in Figure 9.1 shows the transmitted signal. In real applications, a radar sends out many cycles of sinusoidal signals during a pulse and the period of a sinusoidal signal should be much shorter. However, to make the figure more intelligible, let us assume this unrealistic case for demonstration purposes. Assume that 100 ns after the start of a transmitting signal, the radar begins to receive a reflected signal. The reflected signal is shown in the middle plot of Figure 9.1.

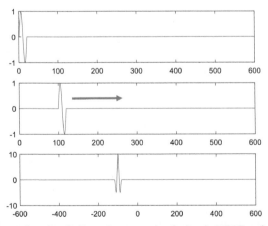

Figure 9.1 Short pulse signal. Top: the transmitted signal. Middle: the received signal reflected from one target. Bottom: the cross-correlation between transmitted and received signals.

Suppose that the radar transmits a pulse every 600 ns (PRI = 600 ns). The top and middle plots of Figure 9.1 illustrate the transmitted and received signals in a window of 600 ns. The radar definitely knows its transmitted signal, so it can calculate the cross-correlation between the transmitted and reflected signal to see if there is indeed a reflected signal. The way to calculate cross-correlation is to first multiply the transmitted and received signal point by point and then sum the product to get one point of cross-correlation. Then, shift the received signal by one sample (in this example, we assume one data point is measured every nanosecond) and repeat the multiplication–summation process to get the next point. Using Figure 9.1 as an example, the transmitted signal remains stationary in this process and the received signal is moving toward either right or left (right is the positive direction indicated by an arrow in the figure and left is the negative direction). The first point of cross-correlation is calculated when the first point of the top plot overlaps with the last point of the middle plot. This process ends when these two plots no longer overlap. By doing so, we get the cross-correlation between transmitted and reflected signals shown in the bottom plot of Figure 9.1. As shown in Figure 9.1, the center index of this cross-correlation is zero and a peak at –100 ns is observed, which means the received signal looks like the transmitted signal delayed by 100 ns. If the radar is stationary, we can assume that the transmitted signal takes 50 ns to reach the target. As the speed of a radio wave is 299,792,458 m/sec, the target is about 15 m away. Of course, this target is extremely close in this demonstration example, which is not realistic. For a moving radar or target, the relative speed between the radar and target can be estimated via Doppler shift and speed needs to be taken into account when determining the target's distance. If there is a target within the radar's detection range, a clear peak with amplitude above detection threshold should appears in the cross-correlation plot.

From the cross-correlation example described above, we can see that the magnitude of cross-correlation is determined by the summation of the transmitted–reflected signal product, and a clear peak happens when the transmitted and reflected signals are aligned. Therefore, to reduce the power of the transmitted signal so that the intercept receiver cannot notice the existence of a radar, a low-power but long PW radar signal can be used. As the total energy of the signal remains the same, radar can still detect the target using cross-correlation. Nevertheless, there is one issue about this approach. If the radar's PW is long, then it is possible that signals reflected from two targets overlap, thus being undistinguishable. Therefore, special radar waveforms need to be used if we desire to transmit a low-power and long-PW signal. This technique is called pulse compression and the radar using these types of waveform is called a pulse compression radar. Both phase coded radar and frequency modulation continuous wave radar will be introduced in this chapter, and we will start with the phase coded radar.

The simplest phase coded radar uses only two types of radar waveforms which are sinusoidal signals with the same frequency and amplitude yet out of phase by $180°$. Let one of them represents 1 and the other represents 0. This signal is referred to as binary phase shifted keying (BPSK). The speed that a BPSK signal can change from 0 to 1 or vice versa is called the chip rate. Instead of sending out one very long sinusoidal signal, phase coded radar sends out cascaded short sinusoidal signals with possible phase changes. For example, a radar can transmit a sequence of BPSK signals as $[1, 1, 1, 1, 1, -1, -1, 1, 1, -1, 1, -11]$ using the two sinusoidal signals just mentioned. This sequence is a 13-bit Barker code. The Barker code is designed to minimize the auto-correlation's (i.e. cross-correlation between a signal and itself) amplitude when two signals are not aligned. To test this design, we replace the one-cycle radar signal with a 13-cycle Barker coded radar signal (PW = 260 ns). The amplitude of the long radar signal is reduced so that the long and short radar signals have the same energy, and the results are shown in Figure 9.2.

Several things can be taken away from Figure 9.2. First of all, despite a much smaller amplitude, the peak amplitude of the cross-correlation of a Barker coded radar signal and its reflected signal shows the same peak

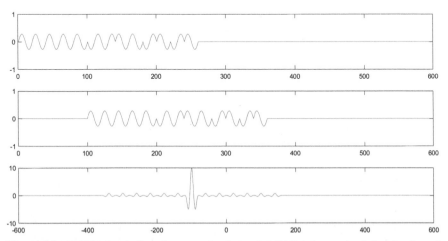

Figure 9.2 BPSK signal. Top: the transmitted signal. Middle: the received signal reflected from one target. Bottom: the cross-correlation between transmitted and received signals.

located at the same index. Therefore, this radar signal can locate the target with sensitivity similar to the pulse radar despite its use of a low power signal. Second, the radar receives the reflected signal even before it finishes transmitting a signal. Therefore, unlike the pulse radar discussed before, this LPI radar needs to use separate antennas, one for transmission and one for receiving. As both antennas are closely placed (so that the receiving antenna can receive reflected signals), the transmission power of this radar cannot be too high and the antenna's sidelobe gain must be much smaller than its mainlobe gain. Otherwise, the receiving antenna will be interfered by the transmitted signal. For this reason, the range of an LPI radar usually is not long. Finally, in this example, despite using signals with PW much longer than the PW of pulse radar signals (260 vs. 20 ns), the cross-correlation of the LPI radar signals displays a peak that appears to be as sharp as the one shown in Figure 9.1. A sharp peak means using this radar signal can still achieve a good resolution to separate targets despite its long PW.

To test if the phase coded radar can separate two targets as good as radar using a short pulse, another simulation is conducted. There are two targets to be detected. For the pulse radar, the signal reflected from the first target reaches the radar 100 ns after the start of transmission and the second reflected signal reaches the radar another 50 ns later. This simulation result is included in Figure 9.3. For the radar using a signal of 20 ns PW, the receiver clearly sees two echoes and two peaks shown in the cross-correlation between transmitted and received signals. For the coded radar whose signal's PW is

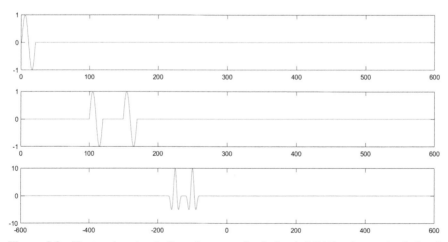

Figure 9.3   Short pulse signal. Top: the transmitted signal. Middle: the received signal reflected by two targets. Bottom: the cross-correlation between transmitted and received signals.

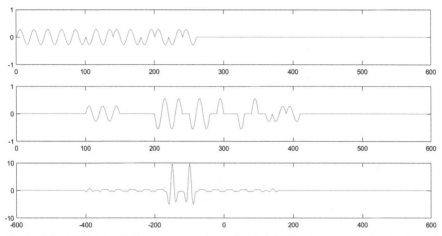

Figure 9.4   BPSK signal. Top: the transmitted signal. Middle: the received signal reflected by two targets. Bottom: the cross-correlation between transmitted and received signals.

260 ns, the two reflected signals overlap, as shown in Figure 9.4. However, two clear peaks at the right places can be observed in the cross-correlation plot. These two simple simulations demonstrate that the coded radar using a lower power yet much longer signal can achieve the same range resolution and sensitivity as the short-pulse radar.

For the EW system, the phase coded radar signal presents a great challenge. Expecting to intercept a radar signal through detecting a sudden received signal power jump will not work. In principle, if the receiver knows the transmitted radar signal, it can apply the same cross-correlation principle to detect the signal, and the energy it can collect might be larger than the energy of the signal received by the radar. However, this is very unlikely the case.

The title of this section is phase coded radar and there is more than one type of phase coded radar. In this section, we only discuss the BPSK radar signal coded with two types of waveforms. There are other phase coded radar signals using more than two waveforms. For example, the quadrature phase shifting keying (QPSK) signal uses four types of waveform with $(0°, 90°, 180°, 270°)$ phase shifts. One issue of the BPSK signal is its vulnerability to Doppler shift, which alters the phase of returned signals. In the next section, another LPI radar waveform, the frequency modulated continuous wave (FMCW), which is more immune to Doppler effects, will be introduced.

9.5 LPI Radar Example 2: FMCW Radar

In Chapter 2, the concepts of FM radar (Section 2.11) and CW radar (Section 2.16) were introduced. The CW radar sends out a non-stop sinusoidal signal and, based on the frequency change of reflected signal, detect a moving target. The issue of a CW radar is that it cannot range the target, and if the relative speed between the target and radar is zero, the CW radar cannot detect the target as there will be no frequency change. The FM radar is a kind of pulse compression radar. Instead of transmitting a strong and short sinusoidal pulse, the FM radar sends out a low-power and long-PW frequency modulated signal. At the receiver side, the long reflected signal can be compressed to reach good range accuracy, as described in Section 2.11, or the cross-correlation between the transmitted and reflected signals can be calculated to get the same effects. An FM signal waveform whose frequency changes from 25 to 125 MHz within 0.25 μs is shown in Figure 9.5. The rate of frequency change is called the chirp rate. The FM signal shown in Figure 9.5 is a linear FM (LFM) signal as the signal frequency changes linearly over time. Other types of FM signal are also used. To demonstrate how cross-correlation can be used to detect targets with FM radar signals, we repeated the two simulations of the previous section. The FM signal shown in Figure 9.5 is used to detect one and two targets. In the case of one target, the radar receiver receives the reflected signal 100 ns after starting the transmitting radar signal. In the case of two targets, two reflected signals are received, 100 and 150 ns after starting transmitting the signal, respectively. The transmitted signal, reflected signals, and the cross-correlation between transmitted

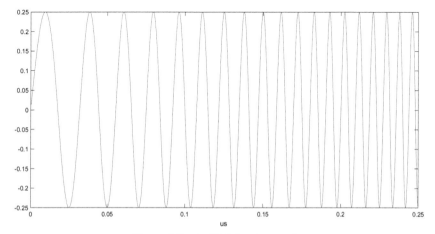

Figure 9.5   An FM signal example.

and reflected signals of two simulations are illustrated in Figures 9.6 and 9.7 respectively. The results demonstrate that a low-power, long-PW FM signal can successfully range the target(s) just like the BPSK signals.

The FMCW radar combines a CW radar and an FM radar by continuously transmitting an FM signal. There are different FMCW signals and the one

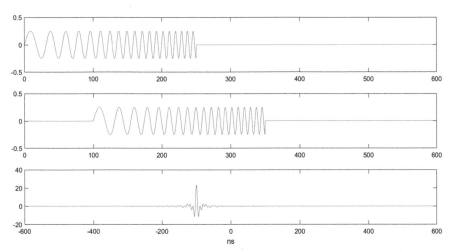

Figure 9.6 FM signal. Top: the transmitted signal. Middle: the received signal reflected by one target. Bottom: the cross-correlation between transmitted and received signals.

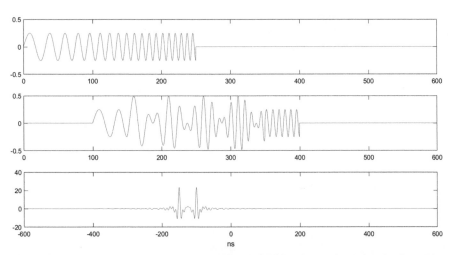

Figure 9.7 FM signal. Top: the transmitted signal. Middle: the received signal reflected by two targets. Bottom: the cross-correlation between transmitted and received signals.

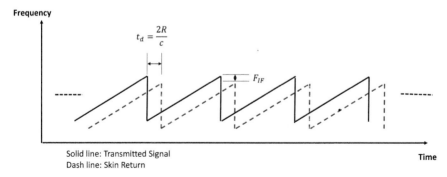

Figure 9.8   Sawtooth FMCW radar signal.

whose time-frequency relation is shown in Figure 9.8 is called a sawtooth signal whose frequency linearly increases until reaching the maximum and then the signal frequency is reset to the starting frequency. The return signal's frequency is different from the transmitted signal's frequency due to the signal's round-trip propagation time (t_d) and Doppler shift (F_{IF}). So, if the receiver calculates the difference between frequencies of transmitted and received signals, during each pulse, there will be two frequency differences. From these two numbers, the target's distance and speed can be determined.

The cross-correlation operation can also be used to range the target in the case of FMCW radar. However, we will cover a widely used FMCW radar diagram, which is a very clever design. This FMCW radar diagram is depicted in Figure 9.9. The FMCW radar has both transmission and receiving antennas. An FMCW signal generated at the receiver is sent to both transmitter antenna and mixer connected to the receiving antenna. The FMCW

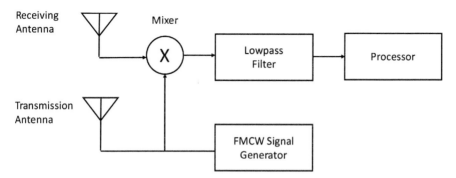

Figure 9.9   FMCW radar diagram.

radar receiver is a superheterodyne (Superhet) receiver, which is covered in Section 4.6. The mixer's inputs are the received signal and the transmitted signal and it outputs two signals whose frequency is the sum of the received and transmitted signal frequencies and the difference between the received and transmitted signal frequencies. The higher frequency component is filtered out by a lowpass filter. So, the resulting signal will be two cascaded sinusoidal signals whose frequencies can be used to locate targets and determine their speeds as described in the previous paragraph.

FMCW radar has been used in a broad spectrum of applications, thanks to its easy implementation. Besides military applications, it is used for security system, speed guns, automotive radar, etc. The sawtooth signal discussed in this section is probably the simplest FMCW signal. Other FMCW signals include a triangular signal whose frequency linearly increases to the maximum value and then linearly decreases back to starting frequency, the frequency shift keying (FSK) whose pulse consists of several short pulses, each of which has different constant frequencies, etc. The interested reader can find a book dedicated to FMCW radars from the reference list [4] at the end of this chapter, but we will stop our discussion on this subject here.

9.6 Countermeasures of LPI Radars

The LPI radar's low signal power makes its detection difficult especially via analyzing a short segment of the intercepted radar signals. In addition, the LPI radar's long duty cycle (in the case of FMCW, the duty cycle is 100%) and low power reduce the chance of the intercept receiver observing signal power fluctuation. So, to detect and classify LPI radar, the intercept receiver needs to analyze a longer segment of received signal with a more advanced signal processing algorithm. Some popular methods for detecting LPI radars such as the Wigner Ville distribution (WVD) or the Choi–Williams distribution (CWD) are designed to analyze how a signal's frequency component changes over time and they are in the category of time-frequency analysis. As this book focuses on fundamental principles and these methods require advanced mathematical concepts to explain, they are not covered in this book.

Not only is the LPI radar difficult to detect, it is also tough to jam. Noise jamming does not work well for two reasons. LPI radars such as the FMCW radar uses a signal with a wider bandwidth, so jamming noise needs to cover a broad bandwidth. Moreover, the LPI radar often uses cross-correlation at the receiver side and there is no similarity between jamming noise and

LPI radar signal. As a result, noise is not an effective jamming tool. If the signal characteristics can be determined, then, to jam a FMCW signal, the EW system can generate another FM signal with a slightly different frequency to provide the wrong speed information. For BPSK radar signals, a BPSK jamming signal can be sent to induce phase error. However, the EW system might not have this information. If the signal's detailed characteristics cannot be extracted or are not available in the database, DRFM might be used to store the intercepted radar waveform, change it slightly, and send it back to deceive the radar. This kind of repeater jamming is more effective than noise jamming.

9.7 Conclusion

LPI radar is understandably a major concern of EW engineers. It is designed to avoid being intercepted, so the radar might detect the aircraft before the intercept receiver on the aircraft detects the radar. Basic LPI radar principles and two popular LPI radar waveforms, BPSK and FMCW, are introduced in this chapter. DRFM can be a useful EA tool to combat LPI radar but only after the EW system is aware of its existence. The next chapter will cover potential applications of machine learning for EW. With rapid advances in machine learning, perhaps it will be the EW engineers' answer to LPI radar.

References

[1] James Genova, *Electronic Warfare Signal Processing*, Artech House, 2017.

[2] R. M. Davis, R. L. Fante, R. P. and Perry, "Phase-Coded Waveforms for Radar," IEEE Transactions on Aerospace and Electronic Systems, vol. 43, no. 1, pp. 401–408, Jan. 2007.

[3] Aytug Denk, *Detection and Jamming Low Probability of Intercept (LPI) Radars*, Master Thesis, Naval Postgraduate School, 2006.

[4] M. Jankiraman, *FMCW Radar Design*, Artech House, 2018.

[5] Jau-Jr Lin, Yuan-Ping Li, Wei-Chiang Hsu, and Ta-Sung Lee, "Design of an FMCW Radar Baseband Signal Processing System for Automotive Application," Springer Plus, vol. 5, Article number: 42, 2016.

[6] Youn-Sik Son, Hyuk-Kee Sung, and Seo Weon Heo, "Automotive Frequency Modulated Continuous Wave Radar Interference Reduction Using Per-Vehicle Chirp Sequences," Sensors, vol. 18, no. 9, Aug. 2018.

[7] Daniel L. Stevens, and Stephanie A. Schuckers, "Low Probability of Intercept Frequency Hopping Signal Characterization Comparison using the Spectrogram and the Scalogram," Global Journal of Researches in Engineering: J General Engineering, vol. 16, no. 2, 2016.

[8] Daniel L. Stevens, and Stephanie A. Schuckers, "Analysis of Low Probability of Intercept Radar Signals Using the Reassignment Method," American Journal of Engineering and Applied Sciences, vol. 8, no. 1, pp. 26–47, 2015.

[9] K. M. Wong, T. Davidson, and S. Abelkader, "Detection of Low Probability of Intercept Radar Signals," Defence R&D Canada – Ottawa Contract Report, DRDC Ottawa CR 2009-142, 2009.

[10] Hung-Ruei Chen, *FMCW Radar Jamming Techniques and Analysis*, Master Thesis, Naval Postgraduate School, 2013.

10

Machine Learning and its Potential in Electronic Warfare

10.1 Introduction

It is difficult, if not impossible, to find a technical field to which no one has tried to apply machine learning (yet). The application of intelligent robots in warfare has been portrayed in movies and TV programs long before machine learning entered our everyday vocabulary. When a technology begins to attract attention, its capability is often overestimated and machine learning is no exception. In this chapter, some possible machine learning applications to EW will be introduced. Unlike some fancy (or very scary) images in people's minds, despite some successful applications, the use of machine learning in EW still has plenty of issues to be overcome, some of which will be discussed in this chapter. However, before jumping into these specific applications, a very loose and informal definition of machine learning will be helpful.

We might naively define machine learning as technology allowing a machine (i.e. a computer) to accomplish tasks through learning from data. In this definition, machines learn how to solve problems by analyzing data rather than reasoning. It is why machine learning and statistic learning are highly related. How well a machine can learn depends on the structure that humans set up in the first place, how the data is presented by humans, and how humans want the machine to "learn." One can classify machine learning based on its "learning style" into different categories: supervised learning, unsupervised learning, semi-supervised learning, and reinforcement learning. Supervised learning means guidance is provided during the learning process. For example, people might use 1000 pictures of dogs and 1000 pictures of cats to train a computer to distinguish between dogs and cats. Each image is labeled with either "cat" or "dog" and the task of the machine is to learn the difference between these two species through provided images so when a new picture appears, the machine can determine whether it is a picture of a dog or a cat. In contrast to supervised learning, for unsupervised learning, the training images fed to machine are unlabeled. Using the dog-cat problem as an example, we might feed 1000 pictures of dogs and 1000 pictures of cats into the machine without telling the machine which picture is that of a dog or that of a cat. Then, the machine's task is to separate the images into two groups, say group A and group B. After the training is over, the machine can label a new picture as a group A picture or a group B picture. One can hope that the two groups that the machine creates are based on the subject's species, but it is not guaranteed; a machine might well separate them based on cuteness of the animal. From this simple explanation, one can say the difference between supervised and unsupervised learning is whether training data is labeled or not. Semi-supervised learning is somewhere between supervised and unsupervised learning. Its training data consists of a small group of labeled data and a large group of unlabeled data. Reinforcement learning differs from the previous learning styles. It does not learn through training data. Instead, it learns how to act in an environment by maximizing the accumulated reward through many simulations. The applications of reinforcement learning include chess playing, computer gaming, etc. In this chapter, we draw examples from supervised and reinforcement learning, mainly due to our personal experiences, so this selection should not be interpreted as unsupervised or semi-supervised learning not having a place in electronic warfare. Also, in this chapter, it is assumed that after the training is complete, the machine's settings remain stationary.

10.2 Machine Learning for Signal Classification

Currently, the most successful application of machine learning for EW applications is arguably signal classification. With rapid advancements in radar technologies, more radar signals have been developed, and to combat advanced radars, correct radar signal classification is crucial. Signal classification is not a new field. It can be used in military or civilian applications. Engineers have developed signal classification methods based on signals' features or a complicated hypothesis testing approach. However, machine-learning-based signal classification methods outperform these traditional techniques. Therefore, it is the first machine learning technique that we discuss.

There is more than one machine learning technique that can be applied to classify a signal. The one introduced in this chapter is the convolutional neural network (CNN). The reason for this choice is that the CNN has been applied to classify LPI radar signals, a formidable threat to EW systems, and has achieved good results. In addition, the CNN is frequently used in introductory machine learning workshops as the first practice example, thus making it a natural starting point for our discussion on machine learning. As the CNN is often used to classify images, to apply the CNN to classify LPI signals, the first task is to convert a one-dimensional signal into an image. Time-frequency signal processing techniques can be used to analyze how the signal's frequency component changes over time and its results can be displayed as an image known as spectrogram. Different signals have different spectrograms and some examples are provided in Figure 10.1. In Figure 10.1, the two axes of the spectrogram are time and frequency, and the color represents intensity. Multiple spectrograms of the same signals can be obtained under different conditions such as noise level, signal strength, time-of-arrival, etc., and these images can be used to train a CNN. After this conversion, a signal classification problem is not much different from distinguishing images of different animals.

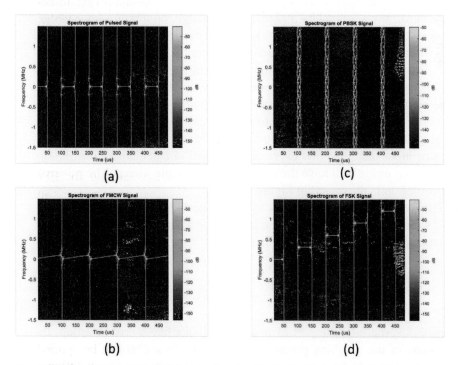

Figure 10.1 Spectrogram of different signals. (a) Short pulse. (b) FMCW. (c) BPSK. (d) FSK.

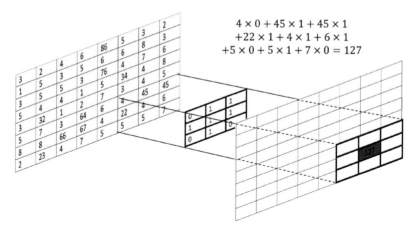

Figure 10.2   Illustration of convolution.

Convolution. To introduce CNN, the notion of convolution must be introduced first. For image processing, convolution can be seen as moving a small
image mask, say a 3 pixel by 3 pixel (3×3) image, on top of a large image.
This image mask is referred to as the kernel. At each position, a pixel-by-pixel
multiplication between the kernel and overlapping portion of the big image is
conducted and all the products (in this case, there are nine of them) are added
together to generate one pixel of resulting image. Then, the kernel is moved
to the next position to repeat the same operation. The number of pixels the
kernel moves each time (referred to as a stride) determines the size of the
resulting image. If the size of a stride is one and zeros are padded around
edges of input image so that every pixel is used in calculating convolution,
the resulting image will have the same size as the input image. The principle
of convolution is illustrated in Figure 10.2.

The purpose of convolution is to extract certain features of the image.
If we look for a cross-pattern in an image, the kernel reassembling a cross
shown below can be used for convolution.

0	1	0
1	1	1
0	1	0

When this kernel and the part of image that looks like a cross overlap, the
magnitude of the resulting product will be large. In a CNN, at each stage,
many kernels might be used in parallel to extract different image features.
As for what kind of features the machine should look for, it is something

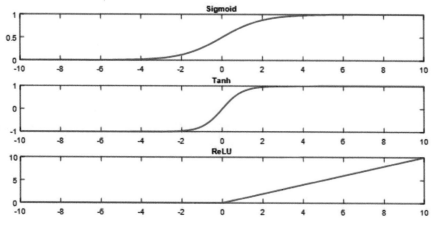

Figure 10.3 Popular activation functions.

the machine needs to learn, and we will discuss this learning process in later paragraphs.

Activation. After the convolution is done, a nonlinear function can be applied to each pixel of resulting image to introduce nonlinearity in the neural network. The function is called the activation function and some types of activation function can be found in Figure 10.3.

Normalization (Bias). Following activation, normalization can be applied to the resulting image to limit the range of its pixel value. One easy way to carry out normalization is to divide the pixel by the average pixel values in its neighborhood.

Downsampling (Pooling). Afterwards, the resulting images need to be downsized to smaller images for further processing. To accomplish this task, an image can be divided into many tiny sub-regions, say a 2×2 sub-image, and replace them with one pixel whose value can be the average or maximum of these four pixels. The layer is called pooling, and a maximum-pooling example is shown in Figure 10.4.

The fours layer described above, *convolution, activation, normalization,* and *pooling,* might be repeated many times (some layers might be skipped in some iterations), and this process is called feature learning. The feature learning stage generates numerous small images representing different features of images. After featuring learning all of the pixels from every image are cascaded into a one-dimensional array (this process is known as flattening) which is the input of neural network for classification.

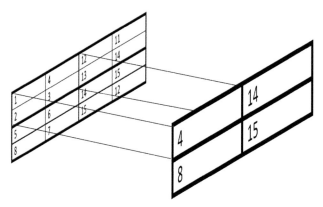

Figure 10.4   Illustration of maximum-pooling.

Artificial Neuron and Neural Network. The neural network got its name from its basic operation unit, the artificial neuron. An artificial neuron and neural network are illustrated in Figure 10.5. The operation of an artificial neuron can be divided into two stages. In the first stage, its inputs are multiplied by weights and then summed together. The value of weights is something the machine needs to learn. Then the sum is used as an input of an activation function and the value of the activation function is the artificial neuron's output. As shown in Figure 10.5, the neural network's first layer is input layer, its last layer is the output layer, and the remaining layers in between are called hidden layers with the artificial neuron as its processing unit. A neural network can have many hidden layers. The phrase deep neural network refers to neural networks with more than two hidden layers. The size of the input layer depends on the number of pixels generated from feature learning. The size of the output layer is determined by the number of signal classes. If the task of the CNN is to distinguish between three radar signals, say pulsed CW, one specific type of FMCW, and one specific type of PBSK, then the size of the output layer is 3. The number of hidden layers and number of artificial neurons in each hidden layer is determined by the programmer. For a classification problem, the output layer's value is usually between 0 and 1. The softmax function defined below is a popular choice for output layer operation:

$$\text{Out}_i = \frac{e^{z_i}}{\sum_{j=1}^{k} e^{z_j}} \tag{10.1}$$

where z_i is the output of the last layer of the artificial neuron. Based on Equation (10.1), the output value must be less than 1. Assume that

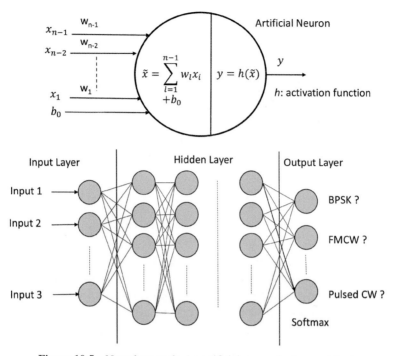

Figure 10.5 Neural network. An artificial neuron is shown at the top.

Out_1 represents the probability of the input signal being a pulsed CW signal, Out_2 represents the probability of input signal being a FMCW signal, and Out_3 represents the probability of input signal being a BPSK signal. In the ideal scenario, when the input is a FMCW signal, a perfect CNN should generate the results of: $Out_1 = 0$, $Out_2 = 1$, and $Out_3 = 0$. The whole CNN diagram can be found in Figure 10.6.

One thing worthy of mention is that it is not absolutely necessary to convert the received signal into an image for classification. The received input signal can be fed into a neural network directly to be classified. In this case, the feature learning portion of a CNN is not used and the size of the neural network's input layer is equals to the length of the sampled received data.

Backpropagation. We have finished our description of a CNN. After the CNN structure is fixed, the performance of this network depends on the value of the kernel and the weight of each neuron. These values are determined in the training phase. The labeled training data helps a machine to determine these coefficients. For example, assume that a pulsed CW is used as training data

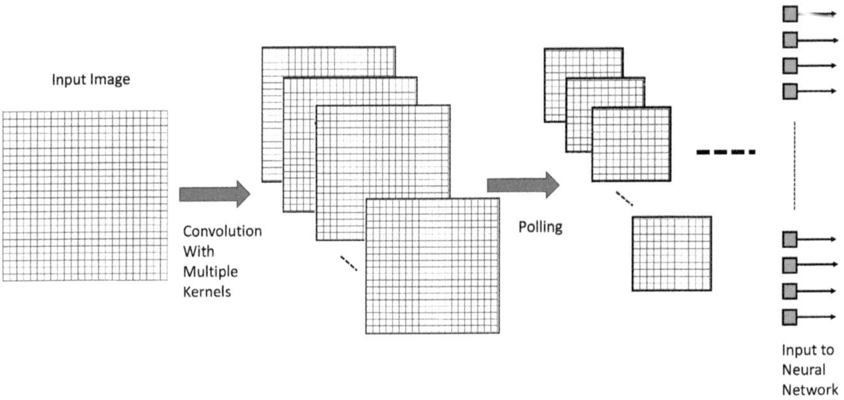

Figure 10.6   Convolutional neural network.

and the CNN network's output is $Out_1 = 0.3$, $Out_2 = 0.6$, and $Out_3 = 0.1$, while the correct result should be $Out_1 = 1$, $Out_2 = 0$, and $Out_3 = 0$. The CNN's coefficients need to be changed based on this information to reduce errors. Back propagation is an algorithm allowing the machine to change the neural network's coefficients to reduce errors, beginning from the last layer toward the first layer by using chain rules and partial derivative. In this process, the error is propagated in a reverse direction to update the CNN's coefficients and it is the reason why this algorithm is called back propagation. The size of adjustment is based on error magnitude and learning rate, a parameter set by the programmer. The hope is that after feeding the CNN with enough training data, the CNN will learn a setting that can deliver an accurate signal classification. One task that remains challenging for engineers working on signal classification is to identify multiple signals arriving at different times with different powers and being partially overlapping in time. One way to deal with this scenario is to train a CNN with as many scenarios an EW receiver might encounter as possible, but this approach can be time-consuming. So, a more efficient solution is needed.

As discussed previously, the number of neural network layers, the size/number of kernels, the number of neurons, learning rate, etc., can affect a CNN's performance and these settings, referred to as hyperparameters, are determined by humans. Also, how well the data is pre-processed also has significant impact on a CNN's performance. An old saying, "garbage in garbage out," accurately captures the essence of a neural network. Unsurprisingly, using machine learning to determine a neural network's hyperparameter is a very active research area. At least, before the determination of hyper-parameters and pre-processing of data for machine learning can be totally done by the machine, people will still be useful, even in fields dominated by machine learning.

10.3 Machine Learning for Predicting a Multifunction Radar's Next Signal

A multifunction radar can change its scan pattern and signal characteristics on a pulse-to-pulse basis following some predetermined rules. To correctly jam such a radar, the EW system needs to be able to predict the radar's next move. To some degree, this task is similar to finishing an incomplete sentence or finishing an incomplete piece of music. Machine learning can be used for this task, but the neural network covered in the previous section is not a suitable approach. The difference between signal classification and this task is the temporal dependence between received signals in the latter case. The CNN treats the received signal as a whole for deciding its type. On the other hand, to predict multifunction radar's next signal, the EW system receives signals in sequence and aims to determine the next signal's type. To handle a sequence of data, a recurrent neural network (RNN) is more suitable than the CNN, and an RNN diagram is shown in Figure 10.7.

As illustrated in Figure 10.7, the RNN has a feedback path. For display purposes, we can unfold this process in time, as shown in Figure 10.7. The RNN's output is generated based on its current input, previous inputs, and previous outputs. One useful concept to explain this dependence is "state." One very loose analogy to this structure can be the following. Treat money a person receives on a Friday as input, the state is their savings, and the output is the amount of money they will spend in the following week. The output and the next state depend on input and current state. In this example, state (savings) is determined by the previous state, input (income),

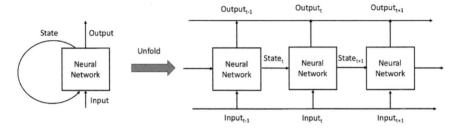

Figure 10.7 Recurrent neural network.

and outputs (spending). The purpose of RNN is to predict the output based on current input and state.

So how can an RNN be trained? Just like CNN, the RNN model consisting of layers of neurons whose structure is set by the programmer and training data is used to determine these coefficients. With training data consisting of input and desired output, the input sequence is fed into the RNN which generates an output sequence. The RNN coefficients can then be modified starting from comparing the last RNN output (rightmost of unrolled RNN, part of which is shown in Figure 10.7) with desired corresponding output and moving toward the first RNN output (leftmost of unrolled RNN). This process is called back propagation through time.

One issue with the RNN is that as the state keeps evolving, the RNN will "forget" things. Despite the state being the accumulation of all of past history, recent events carry more weight. On the other hand, some long-term memory can be useful for processing. Using the previous example again, the expense can be largely determined by savings and income. However, someone winning a lottery at a young age might have spending habits very different from someone being laid-off at a young age. To solve this issue, a special RNN, a long short-term memory (LSTM), can be used. A diagram of LSTM is depicted in Figure 10.8. The LSTM has an additional carry memory (CM) unit which can maintain long-term memory and shape the state. The LSTM output is calculated using state, carry memory, and input. This modification gives LSTM more flexibility.

The RNN has been widely used in natural language processing. To some degree, if we consider every different signal used by a multifunction radar as a specific word, then a sequence of radar signals intercepted and correctly classified is like a sentence (different words can be assigned to different

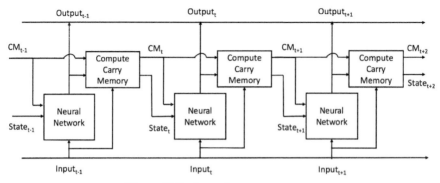

Figure 10.8   Long short-term memory.

signals) and to predict the next signal is like guessing the next word of a half-finished sentence based on the knowledge of a specific grammar. For this kind of application, LSTM might be a valuable tool.

10.4 Reinforcement Learning for EW Actions

In 2016, a British company, DeepMind, achieved something many people suspected could not be done for another generation: developing a computer program, AlphaGo, beating a world-class Go player, 18-time world champion Lee Sedol, in a five-game series. The ancient board game of Go is a strategic game; the player who occupies more territories wins. As the possible configurations in the game of Go are more than 10^{170}, using computers to conduct an exhaustive search for the best move was not yet possible. DeepMind used reinforcement learning to develop AlphaGo and machine learning allows the machine to learn how to play and win a game without learning rules first. If the reinforcement learning can beat best human players in a game of strategy, it is only natural to assume that this technology can be used in EW systems' engagement with radars. In this section, some basic concepts about reinforcement learning are introduced, and potential issues of applying reinforcement learning in EW will also be discussed. However, before we start, some terminology needs to be defined.

- *Agent:* the one taking action in a game.
- *Action:* the decision the agent makes.
- *State:* status of the agent in the environment. The agent might have full (like in the game of Go) or partial (like in the game of Black Jack) knowledge about the state it is in.
- *Reward/Penalty:* quantities defined by the programmer. It can be how many territories a Go move helps to gain (a reward) or how much money a Black Jack player loses (penalty).
- *Policy:* the algorithm the agent uses to choose its action. The purpose of reinforcement learning is to determine the optimal policy to maximize reward.
- *Quality Value (Q-value):* expected discounted reward (minus penalty) after a certain action is taken at a certain state. It can be written as $Q(s, a)$, where s is a state and a is an action. The reward (and penalty) is reduced depending on when it is received after the action. For example, a reward after three steps is multiplied by γ^3, where γ is the discount factor whose value is less than one. In other words, results obtained soon after the action receive higher weight.

It is assumed that an agent is in an environment with multiple states and has several options of actions. The agent does not need to know the rules of the game at the beginning. So, it might take random actions initially to determine values for each state and actions through many simulations (also known as episodes). It is usually assumed that the agent's next state only depends on the state it is currently at, the action taken by the agent, and probability. This setting is referred to as the Markov decision process, one of which can be found in Figure 10.9. In Figure 10.9, big circles stand for state, small circles represent actions, and the number next to line is the state transition probability. In this example, starting from state S_0, if the agent takes action a_0, it has 50% probability of remaining at S_0 and 50% of chance of moving to state S_2. For the agent in state S_2, if it takes action a_1, there is 20% chance that its states will go back to S_0, and when it happens, the agent will receive a reward of 2.

Many simulations are needed to determine the Q-value for each combination of state and actions. After an accurate estimation of the Q-value is obtained, the optimal policy will be simply choosing an action that generates

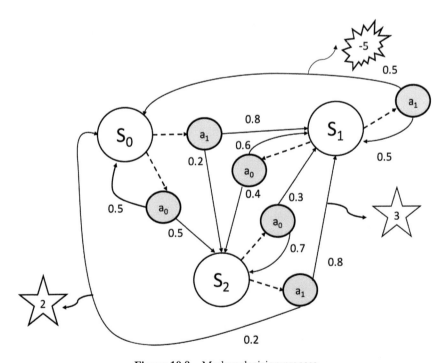

Figure 10.9 Markov decision process.

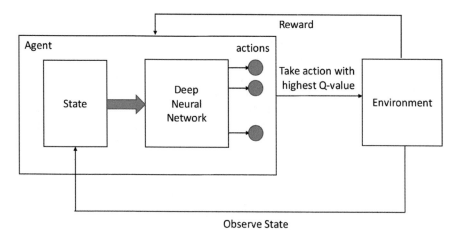

Figure 10.10 Simplified deep-Q learning.

the highest Q-value. However, this approach requires each combination of states and actions to be visited in the training process. This requirement is not realistic when there are a large number of states. To solve this problem, deep-Q learning is proposed.

A simplified deep-Q learning diagram is shown in Figure 10.10. A neural network uses a state as input and as output the Q-value of each possible action. When an action is taken, the reward it generates is used to update the Q-value and train the neural network. After the network is well-trained, in any state, the action with the highest Q-value will be recommended. Although this concept of deep-Q learning is straightforward, it has an inherent issue: the neural network chases a moving target. When the neural network is updated, the calculation of the Q-value is also changed. As a result, deep-Q learning does not always generate a stable outcome and some modifications are necessary. Fixing deep-Q learning is one of DeepMind's major contributions. However, we will stop our discussion about deep-Q learning here before it becomes too complicated for an introductory book.

A policy can be deterministic or stochastic. A deterministic policy means the agent always takes the action that is supposed to produce the biggest reward. The stochastic policy means the agent takes an action based on a predetermined probability. For example, the agent might take an action with the highest Q-value 95% of the time, and the probability of the agent taking other actions is 5%. The reason for stochastic policy is to explore new actions so that the machine can learn some new tricks.

Despite reinforcement learning stunning the world by dominating the human in the game of Go, there are some significant differences between Go and EW. In a game of Go, every piece of information about a current situation is available and any possible move the opponent can make is known. By contrast, for an EW system, not all information is known for certain; the enemy might conceal its intentions, and, worst of all, an EW system does not necessarily know every move its adversary can make and might not have appropriate countermeasures. For example, a new type of missile might be secretly deployed. A reinforcement learning system that keeps learning during the operation (just like a pilot or an EW officer) might alleviate this problem. However, such a system could require a huge number of calculations and the risk of losing it to the enemy's hands could be too tremendous to take. Cloud computation that handles calculations remotely might be a possible solution in this case.

10.5 Possible Issues of Applying Machine Learning for EW

Machine learning has shown great potential or value in many fields and companies, making an early investment on machine learning like NVidia, profits greatly from its forward thinking. We have seen some applications of machine learning in EW and it is guaranteed that more machine learning methods will find their advocates among EW researchers. Nevertheless, there is one major difference between EW and other fields: the lack of training data. For machine learning to perform well, a great amount of training data is needed but radar operators would like to keep their radar signals secret. In the case of reinforcement learning, sufficient episodes need to be played in a realistic environment to train the system well, but a new radar not considered in simulation or a radar whose functionality is not fully understood might make previously learned policies unserviceable. How well machine learning can help EW with a limited amount of information might be questionable. Nevertheless, the issues of a limited amount of training data in other applications such as credit fraud detection (normal transactions significantly outnumber fraudulent transactions) are something machine learning experts have been long aware of. Techniques like generating synthetic training data based on real data might be helpful in training an EW system; however, the verdict is still out.

10.6 Conclusion

This chapter introduces some current and potential applications of machine learning in EW. Machine learning is a rapidly changing field, and this chapter only covers few selected basic machine learning concepts and omits much detail. Readers interested in machine learning can find some accessible references at the end of this chapter and there are many free tutorials and books available online. Besides the applications covered in this chapter, machine learning might be useful in other EW applications like RF component signature identification, anomaly detection, etc. However, something that should be borne in mind is that machine learning can be used to improve both EW systems and radars. So, it will be probable that when one machine learning technology is used by EW engineers to combat radars, engineers on the other side might use the same or a different machine learning technology to fight back.

References

[1] Paul Wilmott, *Machine Learning: an Applied Mathematics Introduction*, Panda Ohana Publishing, 2019.

[2] Francois Chollet, *Deep Learning with Python*, Manning, 2018.

[3] Andriy Burkov, *The Hundred-Page Machine Learning Book*, Self-Published, 2019.

[4] Aurelien Geron, *Hands-On Machine Learning with Scikit-Learn, Keras and TensorFlow*, O'Reilly, 2nd edition, 2019.

[5] Elsayed Elsayed Azzouz and Asoke Kumar Nandi, *Automatic Modulation Recognition of Communication Signals*, Kluwer Academic Publishers, 1996.

[6] O.A. Dobre, A. Abdi, Y. Bar-Ness, and W. Su, "Survey of Automatic Modulation Classification Techniques: Classical Approaches and New Trends," IET Communications, vol. 1, no. 2, pp. 137–156, Apr. 2007.

[7] M. Zhang, M. Diao, and L. Guo, "Convolutional Neural Networks for Automatic Cognitive Radio Waveform Recognition," IEEE Access, vol. 5, pp. 11074–11082, 2017.

[8] S. Kong, M. Kim, L. M. Hoang, and E. Kim, "Automatic LPI Radar Waveform Recognition Using CNN," IEEE Access, vol. 6, pp. 4207–4219, 2018.

[9] N. Visnevski, V. Krishnamurthy, A. Wang, and S. Haykin, "Syntactic Modeling and Signal Processing of Multifunction Radars: A Stochastic Context-Free Grammar Approach," Proceedings of the IEEE, vol. 95, no. 5, pp. 1000–1025, May 2007.

[10] A. Wang and V. Krishnamurthy, "Signal Interpretation of Multifunction Radars: Modeling and Statistical Signal Processing With Stochastic Context Free Grammar," IEEE Transactions on Signal Processing, vol. 56, no. 3, pp. 1106–1119, Mar. 2008.

11

Conclusion

We have reached the end of this journey. Hopefully, you will agree with us that electronic warfare is an interesting technical field and the ever-evolving engagement between radar and EW systems demonstrates mankind's creativity, which can be fascinating and frightening at the same time. Electronic warfare is a direct response to military radar which, thanks to rapid advances in computer processing power and solid state devices, has become more complicated than ever. The development of LPI radars might well eradicate the major advantage of EW systems over radars, as they might no longer be able to detect radars before being detected. The availability of low-cost software-defined radio types of devices and ubiquitous wireless communications networks make more powerful and affordable passive and netted radars possible. To win a war against the huge number of different adversaries in an extremely congested electromagnetic spectrum will be a daunting task for EW engineers. It probably is not a wild guess that DRFM and machine learning will play important roles in this fierce electromagnetic spectrum war.

Due to the authors' research experience, the discussion on EW in this book is limited to air warfare involving aircraft and missiles (i.e. air EW). Maritime EW and ground EW are both interesting fields on which many talented engineers focus. Space EW aiming to protect satellites has become an active research field for good reason. However, as Ludwig Wittgenstein put it, "Whereof one cannot speak, thereof one must be silent," these subjects are not covered in this book due to our lack of knowledge in these fields.

This book is written to introduce fundamental EW concepts along with some battle examples. Therefore, the use of equations is intentionally limited. For readers interested in EW history, late Alfred Price is a prolific author on this subject. Readers who want to learn more technical details about EW or who are considering a career in this field can find a wide selection of engineering books on radar/EW published by Artech House. The authors of this book have also published a college-level EW textbook with IET.

Finally, this book is concluded with a Latin adage attributed to Vegetius: *Si vis pacem, para bellum* (if you want peace, prepare for war).

Index

Electronic countermeasures (ECM) 43, 47, 117
Electronic protection (EP) 47
Electronic support measures (ESM) 40
Electronic warfare (EW) 9, 47, 51, 56, 61, 87, 129
Electronic warfare aircraft 164
Electronic warfare processor 45, 51, 61, 66, 68

F
F-105F 89
F-117 121, 124, 126
F-16 58
F-18 117
F-35 118, 122, 124, 127
F-4 58
Flare 43, 109, 110, 113, 115
Francis Gary Powers 47
Frequency hopping 63, 65, 132, 142
Frequency modulated (FM) 25, 133, 137, 142
Frequency shift keying (FSK) 141

G
George S. Patton, 102
Guglielmo Marconi 1, 3

H
Heinrich Rudolf Hertz 2
Henry Boot 22
Hermann Göring 115
High-speed anti-radiation missile (HARM) 44
Hope diamond 123

I
Infrared (IR) 96, 97, 100, 107, 113
Infrared guidance system 105
Infrared guided (IR) 110
Infrared receivers 96, 99
Intercept receiver 41, 45, 51, 53, 61
Intermediate frequency (IF) 58

J
Jack Monk 115
James Tsui 15, 24, 55, 61, 98, 122
Jamming 1, 8, 18, 47, 51, 71, 85, 113
Noise jamming 74, 77
Barrage jamming 76
Spot jamming 76
Sweep spot jamming 76
Deceptive jamming 46, 78, 82
Sidelobe jammer 82
Inverse gain jamming 84, 131
Cross-eye jamming 85, 86
IR jammer 107
Jennet Conant 22
John Randall 22

K
Knickebein 6, 7

L
Lee Sedol 155
Linear FM (LFM) 138
Long short-term memory (LSTM) 154
Look through window 45
Looking through time 67
Lorenz system 5, 7
LORO (Lobe on Receive Only) 85
Low probability of intercept (LPI) 36, 41, 59, 107

M
Man-portable air defense systems (MANPADS)104
Markov decision 156
Maxwell James Clerk 2
Miniature air-launched decoys (MALD)117
Missile approach warning system 96, 97, 98
Missile guidance systems 104
Mixer 58, 140

O
Operation desert storm 110, 117, 119

About the Authors

Chi-Hao Cheng is a professor with the Department of Electrical and Computer Engineering, Miami University, Oxford, OH, USA. Before joining Miami University, he worked in the optical communications industry for five years. His primary professional interests lie in optical communications, digital signal processing, and electronic warfare receiver development. His research work has led to more than 50 academic articles and his research projects have been sponsored by AFRL, DARPA, NAVAIR, NSF, etc. He is the co-inventor of three US patents.

James Tsui was the world's leading authority on electronic warfare (EW) receiver technology. His pioneering work resulted in over 50 awarded patents, and he was recognized as an Air Force Research Laboratory (AFRL) and Institute of Electrical and Electronics Engineers (IEEE) Fellow. Among his 90 publications, Dr. Tsui authored seven books on receivers: two on analog EW receivers, three on digital EW receivers, and two on software GPS receivers. These books are used as primary references by engineers and scientists developing advanced receiver technologies for many diverse sensing applications, including EW, radar, and communications and navigation systems. The books have been the basis for university graduate courses and highly sought-after short courses developed and taught to all three DoD services and DoD industrial companies internationally. Dr. Tsui passed away in 2019.

For Product Safety Concerns and Information please contact our
EU representative GPSR@taylorandfrancis.com Taylor & Francis
Verlag GmbH, Kaufingerstraße 24, 80331 München, Germany